Zoologica

Original Contributions to Zoology

Founded 1888 by R. Leuckart, C. Cun, continued by W. Kükenthal, R. Hesse, W. E. Ankel

Edited by Thies Büscher, Gerhard Steiner, Klaus-Dieter Klass, Carsten Müller

Volume 167

Bernhard J. van Vondel

Immature stages of Haliplidae. Review of the current knowledge, additions and corrections (Coleoptera)

With 80 plates

Schweizerbart Science Publishers
Stuttgart 2025

Bernhard J. van Vondel

Immature stages of Haliplidae

Authors address:
Bernhard J. van Vondel, Natural History Museum, p/o Roestuin 78, 3343CV Hendrik-Ido-Ambacht, The Netherlands.
peltodytes@gmail.com

This volume is edited by Thies Büscher

Front cover: *Peltodytes intermedius,* provided by M. Hayashi

ISBN 978-3-510-55056-2
ISSN 0044-5088

Information on this title: **www.schweizerbart.com/series/zoologica**

Publisher: E. Schweizerbart'sche Verlagsbuchhandlung (Nägele u. Obermiller),
Johannesstraße 3A, 70176 Stuttgart, Germany
mail@schweizerbart.de
www.schweizerbart.de

∞ Printed on permanent paper conforming to ISO 9706-1994
Printed in Germany by e.kurz + co., Stuttgart

Zoologica
Vol. 167, 1–100
Stuttgart, March 2025

Immature stages of Haliplidae. Review of the current knowledge, additions and corrections (Coleoptera)

Bernhard J. van Vondel

With 80 plates

Abstract

All species of Haliplidae species of which one of the immature stages or biology is described are listed here. Some new descriptions and corrections and additions to earlier descriptions are presented. Most information on larval stages is available from the European species, followed by species from the Nearctic, Australian and Japanese regions, and distinctly less from the Neotropical, Afrotropical and Oriental fauna. The application of suitable collecting methods will likely yield new larval material from previously neglected regions. Lists of phylogenetically relevant characters are presented. Future phylogenetic evaluations of larval features should be encouraged.

Keywords: Haliplidae; eggs; larvae; pupae; immature stages

Contents

www.schweizerbart.de
0044-5088/2025/0167 $ 45.90

Introduction

Haliplidae are a small and unusual family of Adephaga, distinguished from the other groups of the suborder by very large metacoxal plates and larvae specialized on sucking out contents of cells of algae with channeled mandibles (Vondel 2016). At present about 233 species are known (Vondel 2023). The immature stages, i.e. eggs, larval instars 1–3 and pupae, of about 75% are unknown. Schiødte (1864) was the first describing immature stages of Haliplidae, the 3rd instar larva of *Haliplus fulvus* (Fabricius) and *H. variegatus* Sturm, and in 1872 of *Peltodytes caesus* (Duftschmid), followed by Rousseau (1919a, b) with the 3rd instar of *Brychius elevatus* (Panzer). Since then, Bertrand (1923, 1925a, b, 1928a, b, c, 1930, 1931, 1933, 1934, 1942, 1949, 1950, 1951, 1954, 1963, 1966a, b, 1968, 1969, 1972) investigated larvae of Haliplidae and other water beetles intensively. Several authors were successful in rearing species of Haliplidae and different investigations dedicated to the group. Studies dealing with the immature stages of Haliplidae are still scattered and an overall picture was missing so far. In this contribution general information on biology, rearing and conservation is listed. A list of the available identification keys is presented. Al species of which one or more of the immature stages is treated are listed, arranged by Faunal Region. When figures are added to the descriptions, these are for the greater part cited here with the appropriate reference or reproduced when permitted by publisher or author (see Credits on the end of this study). For further details or explanations, the original study should be consulted. For a number of species new or additional descriptions are included here. Figures followed by an * were made by the author for this study and were not published before. Suggestions are made for further research to reach a better understanding of the phylogeny.

General (immature stages)

Rupertsberger (1879) summarized the known European larvae of Haliplidae (4 species at that time).

Matheson (1912) investigated the biology and reared two species of *Peltodytes* Régimbart and one of *Haliplus* Latreille. Details are given there.

Rousseau (1919b) shortly described the larvae of Haliplidae.

DeRyke, W. (1922) recorded larvae of Haliplidae in the stomach of 6 different fish species.

Bertrand (1925a, b) lists the larvae of Haliplidae actually known or recently collected by him (no descriptions).

Bertrand (1928a, b, c) published a comprehensive treatment of immature Haliplidae with a number of new descriptions.

Hickman (1931) reported on the biology of Haliplidae and on rearing methods.

Bertrand (1933) wrote an informative article about larvae.

Bertrand (1934) published a short study on the collecting, raising and conservation of larvae.

Hickman (1937) reported on methods to rear species of Haliplidae.

Hinton (1947) examined spiracles in aquatic larvae of Holometabola, concluding that they are only functional in 3rd instar larvae of Haliplidae, on the mesothorax and the first seven abdominal segments.

Bertrand (1949) reported on the food-plants of the larvae.

Bertrand (1950) reported the capture of additional larvae of a number of species giving information on the food plants. He also mentioned that larvae of *Haliplus* (*Haliplidius*) *varius*, *H.* (*Haliaplus*) *laminatus* and *H.* (*Liaphlus*) *guttatus* are still unknown.

Zaitzev (1953) reported on the biology and described various stages: egg, larvae and pupa (figures are copied from Bertrand 1923/1928b).

Bertrand (1954) presented a chapter on Haliplidae in "Les Insectes aquatiques d'Europe" with a key to the pupae of *Haliplus* and *Peltodytes*. He also illustrated the two types of foreleg of species of *Haliplus*: with or without a "thumb" or "grasping device" (pouce in French) on the tibia, but he switched the figures 1 and 2.

Jaboulet (1960) intensively studied, described and illustrated the [3rd instar] larvae of *Brychius* Thomson, *Haliplus* and *Peltodytes*. In the case of *Haliplus* however, not all details apply to all species like the shape of the forelegs (Plate 1–3).

Peterson (1960) gave a general description of the family and the genera *Haliplus* and *Peltodytes*.

Seeger (1971a, b, c) intensively studied the biology of a number of species and described eggs, larvae and pupa.

Bertrand (1972) summarized the current knowledge on haliplid larvae in "Larves et Nymphes des Coléoptères Aquatiques du Globe" and provided a key to the genera.

Beutel (1986) published a detailed account of the head anatomy of *Haliplus lineatocollis*.
Spangler (1991) presented a general description of the family and North American genera.
Wichard et al., (1995) investigated the feeding and breathing of larvae of Haliplidae.
Kitano & Sano (2011) observed larvae of Haliplidae feeding on algae.
Makarov & Prokin (2015) studied the postanal process of larvae of *Haliplus* and concluded that the often apically forked terminal segment (by some authors regarded as partly fused urogomphi or cerci) should be considered as outgrowth of the dorsal surface of segment X, not homologous to urogomphi or cerci as interpreted by Vondel (2004, 2005, 2011; Beutel et al. 2011; Jaboulet 1960).
Linhart et al. (2023) claimed that they found larvae in Myanmar amber probably belonging to the beetle family Haliplidae, but Beutel & Vondel (2024) showed that they belong to the polyphagan tenebrionoid family Scraptiidae (subfamily Scraptiinae).

Identification keys (larvae and pupae)

Rousseau (1919a, b) provided a key and short description (illustrated) to the 3rd instar larva actually known (*Brychius elevatus, Haliplus ruficollis*(?)*, H. fulvus, H. variegatus, Peltodytes caesus*).
Bertrand (1928b) presented a key to the 3rd instar larvae of ten Central European species and to three pupae (*Peltodytes*, *Haliplus* group 1 and *Haliplus* group 2).
Böving & Craighead (1930) provided a key to the subfamilies Haliplinae and Peltodytinae, but they are the only authors so far recognizing subfamilies.
Henriksen (1930) published a key to about half of the Danish species [in Danish].
Pennak (1953) presented a key to the North American genera (3rd instar larvae).
Zaitzev (1953) published a key to the genera *Brychius*, *Haliplus* and *Peltodytes* and to twelve European species of *Haliplus* (3rd instar larvae).
Leech & Chandler (1956) presented keys to the 3rd instar larvae of Californian Haliplidae.
Bertrand (1972) included a key to the genera *Apteraliplus* [according to Vondel (2019) a subgenus of *Haliplus*, confirming the view of Beutel & Ruhnau (1990) that *Algophilus* and *Apteraliplus* are nested within this genus], *Brychius*, *Haliplus* and *Peltodytes* and to the pupae of *Haliplus* and *Peltodytes*.
Klausnitzer (1977, 1984) presented a key to European species of *Brychius* Thomson, *Haliplus* Latreille and *Peltodytes* Régimbart (figures obviously copied from earlier studies).
Rozkošný (1980) provided a key to the 3rd instar larvae of 13 Polish species using figures from earlier authors [in Polish].
Holmen (1987) presented a key for the 3rd instar larvae of Central European species of genera and subgenera.
Klausnitzer (1991) provided a key to most of the Central European species.
Klausnitzer (1996) published a key to 3rd stage larvae of the genera *Brychius*, *Haliplus* and *Peltodytes*.
Vondel (1997) presented a key for the genera *Brychius*, *Haliplus* and *Peltodytes*, a key for distinguishing the 1st, 2nd and 3rd instar larvae, and a key to the 3rd instar larvae of 18 European *Haliplus* species.
Vondel (2011) published a key to the known 3rd instar larvae of Nearctic Haliplidae l (ca 12 species).

Palaearctic Region

Brychius elevatus (Panzer)

Plate 4–5.
Rousseau (1919a, b) described and illustrated the 3rd instar larva.
Bertrand (1925a, b) reports the capture of larvae.
Bertrand (1928b) presented an illustrated description of 1st, 2nd and 3rd instar larvae.
Zaitzev (1953) published descriptions of the 3rd instar larvae, with drawings of the body and abdominal tergite I (obviously copied from Bertrand 1928b).
Jaboulet (1960) described and illustrated in detail the 3rd instar larva of *Brychius* [*elevatus*].
Wichard et al. (1995) showed SEM micrographs of various parts of the 3rd instar larva.
Beutel (1997) presented a picture of the head of the 3rd instar larva.
Vondel (1997) published some images of the 3rd instar larva.
Vondel (2016) showed SEM pictures of the larval head and of the microtracheal gills of the 3rd instar larva. The 2nd instar larva is described in the chapter “New and additional figures and descriptions”.
Measurements: total length, 1st instar: 3 mm, 2nd instar: 4–6.5 mm, 3rd instar: 8–11 mm (Rousseau 1919a, Bertrand 1928b, this study).

*Haliplus (Haliaplus) laminatus (*Schaller)

Plate 6–7.
Holmen (1981) illustrated the foreleg and mandible of the 3rd instar larva.
Vondel (1986) described and illustrated the 2nd and 3rd instar larva.
Vondel (1997) presented some pictures of the 3rd instar larva.
Measurements: total length, 2nd instar: 4.0–4.2 mm, 3rd instar: 6.5–7.8 mm (Vondel 1986).

Haliplus (Haliplidius) confinis Stephens

Plate 8–11.
Schiødte (1864) described and illustrated the 3rd instar of *H. ruficollis*, but Vondel (1997) stated that it was in fact the larva of *H. confinis*. Comparison with the figures of Seeger (1971a) confirm that assumption.
Bertrand (1928b) described and illustrated the 3rd instar larva and pupa.
Böving & Craighead (1930) illustrated the body, head and hindleg.
Zaitzev (1953) published a drawing of abdominal tergite I of the 3rd instar larva (copied from Bertrand 1928b).
Seeger (1971a) presented an illustrated description of the 3rd instar larva and the pupa.
Vondel (1997) published some pictures of the 3rd instar larva. In this study the 1st and 2nd instar larvae are described in the chapter "New and additional figures and descriptions".
Measurements: total length, 1st instar: 2.5 mm, 2nd instar: 5 mm, 3rd instar: 8.1–9.7 mm, pupa: 2.75–3.5 mm (Bertrand 1928b, Seeger 1971a, this study).

Haliplus (Haliplidius) obliquus (Fabricius)

Plate 12, 13.
Bertrand (1928b) described and illustrated the 2nd and 3rd instar larva and pupa, but reported the discovery earlier (1925a, b).
Zaitzev (1953): presented a drawing of the first abdominal tergite of the 3rd instar larva (obviously copied from Bertrand 1928b).
Seeger (1971a): presented an illustrated description of the 3rd instar larva and the pupa.
Vondel (1997) presented some pictures of the 3rd instar larva.
In this study the 2nd instar larvae is described in the chapter "New and additional figures and descriptions".
Measurements: total length: 2nd instar: 5–5.5 mm, 3rd instar: 8–9.7 mm, pupa: 2.9–3.5 mm (Bertrand 1928b, Seeger 1971a, this study).

Haliplus (Haliplidius) varius Nicolai

Plate 14, 15.
Vondel (1996) described the 3rd instar larva.
Vondel (1997) presented some pictures of the 3rd instar larva. He also discussed the subgeneric status.
In this study figures of the 2nd instar larva are presented in the chapter "New and additional figures and descriptions".
Measurements: total length: 2nd instar: 2.8 mm, 3rd instar: 7.5–8.1 mm (Vondel 1996, this study).

Haliplus (Haliplus?) sp.

Plate 16.
Kitano & Sano (2011) published a photograph of a 2nd (?) and 3rd instar larva.

Haliplus (Haliplus) apicalis Thomson

Plate 17.
Bertrand (1942) described and illustrated the 3rd instar larva.
Vondel (1997) presented some pictures of the 1st, 2nd and 3rd instar larvae.
Measurements: total length: 1st instar: 1.7–1.8 mm, 2nd instar: 3.4–3.7 mm, 3rd instar: 6.3–8.0 mm (Bertrand 1942, Vondel, own observation).

Haliplus (Haliplus) fluviatilis Aubé

Plate 18, 19.
Gernet (1868) described the larva of *H. lineolatus*, but Bertrand (1923/1924) considered it as *H. fluviatilis*. Gernets description is too limited to decide which of the species from the "*ruficollis*"group he was actually dealing with.
Bertrand (1950) described and depicted the 3rd instar larva.
Vondel (1997) presented some pictures of the 3rd instar larva.
In this study the 2nd instar larvae is described in the chapter "New and additional figures and descriptions".
Measurements: total length: 2nd instar: 5.2–5.5 mm, 3rd instar: 6.8–9 mm (Bertrand 1950, this study).

Haliplus (Haliplus) fulvicollis Erichson

Plate 20.
Holmen (1987) showed the first abdominal segment of the 3rd instar larva.
Vondel (1997) presented a picture of the pronotum of the 3rd instar larva.
In this study more figures of the 3rd instar larva are presented in the chapter "New and additional figures and descriptions".
Measurements: total length: 3rd instar: 6.2–7.8 mm (this study).

Haliplus (Haliplus) furcatus Seidlitz

Plate 21.
Vondel (1997) presented pictures of eighth and ninth abdominal segment and the mandible of the 3rd instar larva.
In this study more figures of the 3rd instar larva are presented in the chapter "New and additional figures and descriptions".
Measurements: total length: 3rd instar: 6.3–6.6 mm (this study).

Haliplus (Haliplus) heydeni Wehncke

Plate 22–24.
Bertrand (1942) described and illustrated the 3rd instar larva and described the 2nd instar larva.
Vondel (1997) presented some pictures of the 3rd instar larva.
In this study the 1st, 2nd and 3rd instar larvae are described in the chapter "New and additional figures and descriptions".
Measurements: total length: 1st instar: 2.0–2.2 mm, 2nd instar: 4.1–4.8 mm, 3rd instar: 6.3–8.0 mm (Bertrand 1942, this study).

Haliplus (Haliplus) immaculatus Gerhardt

Plate 25–27.
Falkenström (1926) has raised larvae and the pupa. He described the life cycle, stating that larvae as well as adults are feeding on plants and that adults of this species as well as other species of Haliplidae consume animal food only in the case of starving. He described the egg, the 1st instar larva (illustrated), the 2nd instar larva and the 3rd instar larva (illustrated).
Bertrand (1950) described and figured the 3rd instar larva.
Seeger (1971a) presented an illustrated description of the 3rd instar larva and the pupa and some information about the eggs.
Vondel (1997) presented some pictures of the 3rd instar larva and the pupa.
In this study the 1st, 2nd and 3rd instar larvae are described or additional figures are presented in the chapter "New and additional figures and descriptions".
Measurements: total length: egg: 0.45–0.53 mm, 1st instar: 1.0–3.3 mm, 2nd instar: 3.8–4.5 mm, 3rd instar: 5.9–10.3 mm, pupa: 2.6–3.0 mm (Falkenström 1926, Bertrand 1950, Seeger 1971a, this study).

Haliplus (Haliplus) kamiyai Nakane

Plate 52. Figs. 3–8
Watanabe & Yamasaki (2020) described the 1st, 2nd and 3rd instar larva and depicted the 3rd instar larva and the pupa.
Measurements: total length: 1st instar: 3.19–3.57 mm, 2nd instar: 5.84–7.68 mm, 3rd instar: 10.3–14.28 mm, pupa: ? mm (Watanabe & Yamasaki 2020).

Haliplus (Haliplus) lineolatus Mannerheim

Plate 28–30.
Gernet (1868) described the 3rd instar larva but was not sure that it was this species. According to Bertrand (1923) it was probably *H. fluviatilis*, but Rousseau (1919b) and Vondel (1997) assume that it might belong to a number of species in the "*ruficollis*" group.
Bertrand (1931) described the 3rd instar larva (with a figure of the mandible and a simple drawing of an abdominal tergite).
Seeger (1971a) presented an illustrated description of the 3rd instar larva and the pupa and some information about the eggs.
Vondel (1997) presented some pictures of the 3rd instar larva.
In this study the 1st, 2nd and 3rd instar larvae are described or additional figures are presented in the chapter "New and additional figures and descriptions".
Measurements: total length: 1st instar: 1.7–2.5 mm, 2nd instar: 3.8–5.1mm, 3rd instar: 6.4–8.5 mm, pupa: 2.47–2.69 mm (Seeger 1971a, this study).

Haliplus (Haliplus) ruficollis (De Geer)

Plate 31–33.
Schiødte (1864) described and illustrated the 3rd instar larva in detail, but Vondel (1997) states that in fact it is *H. confinis*.

Rousseau (1919b) presented pictures of the 3rd instar larva and its foreleg, although he placed a ? after the species name.
Bertrand (1923) discussed and illustrated the antenna, mandible, abdominal segments V and X, all of the 3rd instar larva, furthermore the last four abdominal segments of the 1st, 2nd and 3rd instar larva. He also showed an abdominal segment of the 2rd and 3nd instar larva, together with the forked end of abdominal segment X of the 1st instar larva.
Bertrand (1925a, b) reports the capture of larvae.
Falkenström (1926) examined the feeding habits, confirming the discovery of Matheson (1912) that larvae are algophagous and not carnivorous as assumed earlier by Schiødte (1872).
Bertrand (1928b) described and illustrated the 1st, 2nd and 3rd instar larva and pupa.
Zaitzev (1953): presented a drawing of the entire body and abdominal tergite I of the 3rd instar larva (copied from Bertrand 1928b).
In this study the 1st, 2nd and 3rd instar larvae are described and additional figures are presented in the chapter "New and additional figures and descriptions".
Measurements: total length: 1st instar: 1.8–2.3 mm, 2nd instar: 3.6–5.4 mm, 3rd instar: 6.8–9.5 mm, pupa: 2.5 mm (Bertrand 1928b, this study).

Haliplus (Haliplus) sibiricus Motschulsky

Plate 34–36.
Beier (1929) extensively examined the biology of *H. wehnckei* Gerhardt (jun. syn. of *H. sibiricus*). He described the egg and the 1st and 3rd instar larvae and also reared at the same time *H. lineatocollis*. Vondel (1997) suggested that in some cases his figures, especially of the body and the mandible, have much in common with the latter species.
Vondel (1997) presented some pictures of the 3rd instar larva.
In this study the 1st, 2nd and 3rd instar larvae are described or additional figures are presented in the chapter "New and additional figures and descriptions".
Measurements: total length: 1st instar: 1.8–2.5 mm, 2nd instar: 3.7–5.3 mm, 3rd instar: 6.6–7.2 mm (this study).

Haliplus (Haliplus) zacharenkoi Gramma & Prisnyi

Plate 37. Fig. 1.
Dyadichko (2009) showed a photograph of the 3rd instar larva.
Measurements: total length: 3rd instar: ? mm.

Haliplus (Liaphlus) abbreviatus Wehncke

Plate 38–40.
Vondel (2011a) described and illustrated the 3rd instar larva of *Haliplus kulleri*. This species was later synonymised with *H. abbreviatus* (Vondel & Litovkin 2017). The dorsal habitus picture in Vondel (2011) is not correct, showing one abdominal segment too much (noticed by Dr. Shcherbakov, Moscow, 2014, pers. comm.). The correct picture is shown in Plate 40, Fig. 1*. Vondel (2011a) also described a 1st and a 2nd instar larvae, but could not determine whether they belong to *Haliplus kulleri* (now *H. abbreviatus*) or *H. maculatus*.
Measurements: total length: 1st instar: 1.5–1.7 mm, 2nd instar: 3.1–3.9 mm, 3rd instar: 7.7–9.3 mm (Vondel 2011a).

Haliplus (Liaphlus) basinotatus Zimmermann

Plate 37. Fig. 3.
Mitamura et al. 2017 showed a photograph of the 3rd instar larva (under the name of *H. basinotatus latiusculus*, which is a jun. syn. of *H. basinotatus*).
Measurements: total length: 3rd instar: 10.0 mm (Mitamura et al. 2017).

Haliplus (Liaphlus) eximius Clark

Plate 37, Fig. 4, 5.
Yamazaki et al. (2020) showed a small photograph of the 3rd instar larva and the pupa.
Measurements: total length: 3rd instar: ? mm, pupa: ? mm.

Haliplus (Liaphlus) flavicollis Sturm

Plate 41, 42
Bertrand (1928b) described the 2nd and 3rd instar larva and the pupa. He reported the discovery earlier (1925a, b)
Zaitzev (1953) presented a drawing of abdominal tergite I of the 3rd instar larva (obviously copied from Bertrand 1928).
Seeger (1971a): provided a description of the 3rd

instar larva and pupa with pictures of the mandible and of the pupa. Some information about the eggs is presented. It should be noted that the mandible apparently belongs to *H. fulvus*, suggesting that the entire description refers to *H. fulvus*.
Vondel (1997) presented some pictures of the 3rd instar larva.
In this study the 2nd and 3rd instar larvae are described and additional figures are presented in the chapter "New and additional figures and descriptions".
Measurements: total length: 2nd instar: 3.5–7.0 mm, 3rd instar: 7.5–12.5 mm, pupa: 3.3–5.0 mm (Bertrand 1928b, Seeger 1971a, this study).

Haliplus (Liaphlus) fulvus (Fabricius)

Plate 43.
Schiødte (1864) provided a detailed description and illustrations of the 3rd instar larva.
Rousseau (1919b) figured the 3rd instar larva.
Bertrand (1928b) described and illustrated the 3rd instar larva and pupa, but reported the find by Leprieur (1887) earlier (1925a, b).
Zaitzev (1953): provided drawings of the body and abdominal tergite I of the 3rd instar larva (obviously copied from Bertrand 1928).
Seeger (1971a) described the 3rd instar larva of *H. flavicollis*, but in fact it was likely *H. fulvus*.
Holmen (1981) illustrated the foreleg and mandible of the 3rd instar larva.
Vondel (1997) presented some pictures of the 3rd instar larva.
Measurements: total length: 3rd instar: 12–13 mm, pupa: 4.25 mm (Bertrand 1928b).

Haliplus (Liaphlus) maculatus Motschulsky

Plate 38, 39, 44.
Vondel (2011a) described and illustrated the 3rd instar larva. He also described a 1st and 2nd instar larvae, but could not determine whether they belong to *Haliplus kulleri* (now *H. abbreviatus*) or *H. maculatus*.
Measurements: total length: 1st instar: [?] 1.5–1.7 mm, 2nd instar: [?] 3.1–3.9 mm, 3rd instar: 11.0–12.0 mm (Vondel 2011a).

Haliplus (Liaphlus) mucronatus Stephens

Plate 45. Fig. 1, 6–9.
Bertrand (1928a) described the 3rd instar larva. The dorsal side is figured.
Bertrand (1928b) described (erroneously as *H. guttatus*) the 1st, 2nd and 3rd instar larva.
Vondel (1997) presented some pictures of the 3rd instar larva.
Measurements: total length: 1st instar: 2.75 mm, 2nd instar: 5–5.75 mm, 3rd instar: 12 mm (Bertrand 1928b).

Haliplus (Liaphlus) ovalis Sharp

Plate 45. Figs. 2–4.
Hayashi (2011) showed a small photograph of the 3rd instar larva in dorsal and lateral view. Sato & Yoshitomi (2018) presented a small drawing of the 3rd instar larva.
Measurements: total length: 3rd instar: 10 mm (Hayashi 2011).

Haliplus (Liaphlus) rubidus Perris

Plate 45. Figs. 5, 10.
Bertrand (1931) described the 3rd instar larva (with figures of the mandible and the last three abdominal tergites).
Measurements: total length: 3rd instar: 12–13 mm (Bertrand 1931).

Haliplus (Liaphlus) sharpi Wehncke

Plate 46.
Nakanishi (2012) reared some larvae and showed photos of the 2nd and 3rd instar larva in dorsal view and the pupa in dorsal, lateral and ventral view.
Mitamura et al. (2017) showed a photo of the 3rd instar larva.
Measurements: total length: 2nd instar: 5 mm, 3rd instar: 10.0–10.3 mm, pupa: 3.2 mm (Nakanishi 2012, Mitamura et al. 2017).

Haliplus (Liaphlus) variegatus Sturm

Plate 47.
Schiødte (1864) provided a detailed description and illustrations of the 3rd instar larva and pupa.
Bertrand (1925a, b) reported the capture of larvae.
Bertrand (1928b) described and illustrated the 1st, 2nd and 3rd instar larva and pupa, but reported the discovery earlier (1925a).
Zaitzev (1953) presented drawings of the first abdominal tergite of the 3rd instar larva (copied from Bertrand 1928b).
Holmen (1987) illustrated abdominal segment I of the 3rd instar larvae.

Vondel (1997) presented some pictures of the 3rd instar larva.
Measurements: total length: 1st instar: 2.35 mm, 2nd instar: 4.4 mm, 3rd instar: 8–11 mm, pupa: 3.5 mm (Bertrand 1928b).

Haliplus (Neohaliplus) lineatocollis (Marsham)

Plate 48–51.
Bertrand (1923) extensively described and illustrated the 1st, 2nd and 3rd instar larva and the pupa (as *H. lineaticollis*).
Bertrand (1925a, b) reported the capture of larvae.
Bertrand (1928b) described and and showed images of the 1st, 2nd and 3rd instar larva and the pupa.
Beier (1929) extensively examined the biology of *H. wehnckei* Gerhardt (jun. syn. of *H. sibiricus*). He described the egg and the 1st and 3rd instar larvae. Beier reared at the same time *H. lineatocollis* and Vondel (1997) suggests that in some cases his figures, especially of the postcephalic body and the mandible have much in common with the latter species.
Zaitzev (1953) presented drawings of abdominal tergite I of the 3rd instar larva (obviously copied from Bertrand 1928).
Seeger (1971a) presented an extensively illustrated description of all three larval instars.
Beutel (1986) described the skeleton, muscles and digestive tract of the head of the 3rd instar larva.
Vondel (1997) presented some pictures of the 3rd instar larva.
Measurements: total length: 1st instar: 1.0–3.0 mm, 2nd instar: 3.5–4.6 mm, 3rd instar: 6.0–9.0 mm, pupa: 3–3.1 mm (Bertrand 1923, 1928b, Seeger 1971a).

Haliplus (Nipponiplus) japonicus Sharp

Plate 52. Figs. 1, 2.
Mitamura et al. (2017) showed a photograph of the 3rd instar larva.
Nakajima et al. (2020) showed a photograph of the 3rd instar larva.
Measurements: 3rd instar: 8.0 mm (Mitamura et al. 2017).

Peltodytes (Peltodytes) caesus (Duftschmid)

Plate 53–55.
Schiødte (1872) provided a detailed description and illustrations of the 3rd instar larva.
Bertrand (1925a, b) reported the capture of several larval instars.
Bertrand (1928b) provided a detailed and illustrated description of the 1st, 2nd and 3rd instar larva and the pupa.
Bertrand (1930) reports the capture of a number of larvae.
Zaitzev (1953) provided an extensive description of the 3rd instar larvae and the pupa, and a drawing of the 3rd instar larva (obviously copied from Bertrand 1928).
Jaboulet (1960) described and illustrated in detail the 3rd instar larva of *Peltodytes* [*caesus*].
Steinmann & Zombori (1984) presented a picture of an antenna, allegedly of *Peltodytes* sp. [*caesus*?]. However, it clearly differs from figures in the other sources.
Holmen (1987) illustrated abdominal segment I of the 3rd instar larva.
Vondel (1997) presented some pictures of the 3rd instar larva.
Measurements: total length: 1st instar: 1.75–2.25 mm, 2nd instar: 3.5–4.5 mm, 3rd instar: 5 mm, pupa: 3.5 mm (Bertrand 1928 b).

Peltodytes (Peltodytes) intermedius (Sharp)

Plate 56. Figs. 1–8.
Fukuda et al. (1959) presented a drawing of the 3rd instar larva and a description in Japanese.
Hayashi (2011) showed a small photograph of the 3rd instar larva in lateral view and the pupa in ventral view.
Mitamura et al. 2017 showed a photograph of the 3rd instar larva.
Nakajima et al. (2020) showed photographs of the eggs, the 3rd instar larva and the pupa.
Hayashi (pers. comm.) sent me photographs of the 3rd instar larva and the pupa.
Measurements: total length: 3rd instar: 7.0–8.0 mm, pupa: 3.6 mm (Fukuda et al. 1959, Hayashi 2011, Mitamura et al. 2017).

Peltodytes (Peltodytes) rotundatus (Aubé)

Plate 55. Fig. 3.
Bertrand (1951) briefly described the 3rd instar larvae and illustrated the difference in the shape of the basal tubercle on the filamentous tracheal gills of *P. caesus* and *P. rotundatus*.
Measurements: 3rd instar: ? mm.

Peltodytes (Peltodytes) sinensis (Hope)

Plate 56. Fig. 9.
Jia et all. (2024) showed a photo of the 3rd instar larvae. Co-author and photographer Zuqi Mai sent me the original photo.
Measurements: 3rd instar: ? mm.

Nearctic Region

Brychius hornii Crotch

Plate 57.
Mousseau (2004) and Mousseau & Roughly (2007) described and illustrated the 2nd and 3rd instar larvae.
Measurements: total length: 2nd instar: 5.0 mm, 3rd instar: 10.0–11.0 mm (Mousseau 2004).

Brychius hungerfordi Spangler

Plate 58.
Strand & Spangler (1994) described the 3rd instar larva, illustrated by a series of SEM micrographs.
Mousseau (2004) and Mousseau & Roughly (2007) described and figured the 3rd instar larva.
Measurements: total length: 3rd instar: 11.5–13.0 mm (Mousseau 2004, Strand & Spangler 1994).

Haliplus (Apteraliplus) parvulus (Roberts)

Plate 59.
Leech & Chandler (1956) presented a key to the larvae of Nearctic *Apteraliplus*, *Brychius*, *Haliplus* and *Peltodytes*. For *Apteraliplus* they gave the following characters: "Third segment of front leg produced and edged by 2 blunt teeth; 9th abdominal segment unforked (except in 1st instar); body with conspicuous spines at lateral edges". No figures were added.
In this study the 3rd instar larvae is described in the chapter "New and additional figures and descriptions."
Measurements: total length: 3rd instar: 5.3–5.9 mm (this study).

Haliplus (Haliplus) immaculicollis Harris

Plate 60, 61.
Matheson (1912) described the egg, 1st and 3rd instar larva.
Wilson (1924) described the 3rd instar larva.
Hickman (1930) described and illustrated the egg, the 1st, 2nd and 3rd instar larva and the pupa.
Hickman (1931) observed the feeding habits of the larvae.
Measurements: total length: egg: 0.34–0.39 mm, 1st instar: 1.4–3.2 mm, 2nd instar: 3.3–4.4 mm, 3rd instar: 7.4–9.0 mm, pupa: 2.07–2.12 mm (Matheson 1912, Hickman 1930).

Haliplus (Liaphlus) cribrarius LeConte

Plate 62.
Hickman (1930) described and illustrated the 2nd and 3rd instar larva and described the pupa. Hickman (1931) observed the feeding habits of the larvae.
Measurements: total length: 2nd instar: 4.3–5.6 mm, 3rd instar: 11.3–12.5 mm, pupa: 3.72–3.81 mm (Hickman 1930).

Haliplus (Paraliaphlus) deceptus Matheson

Plate 63.
Vondel (2011 b) described and illustrated the 3rd instar larva under the name of *Haliplus variomaculatus* Brigham & Sanderson.
Vondel (2023) concluded that this species is a jun. syn. of *Haliplus deceptus*.
Measurements: total length: 3rd instar: 8.6 mm (Vondel 2011b).

Haliplus (Paraliaphlus) triopsis Say

Plate 64.
Hickman (1930) described and illustrated the 2nd and 3rd instar larva and the pupa.
Hickman (1931) observed the feeding habits of the larvae.
Measurements: total length: 2nd instar: 5.46–5.88 mm, 3rd instar: 5.88–13.44 mm, pupa: 3.27–3.4 mm (Hickman 1930).

Peltodytes (Neopeltodytes) duodecimpunctus (Say)

Plate 65, 66.
Hickman (1930) described (as *P. lengi*) and illustrated the egg, the 1st, 2nd and 3rd instar larva and the pupa.
Hickman (1931) observed copulation, egg-laying and feeding habits of the larvae.
Chandler (2008) presented a photograph of the 3rd instar larva.
Measurements: total length: egg: 0.415–0.483 mm, 1st instar: 0.65–0.8 mm, 2nd instar: 2.8–4.2 mm, 3rd instar: 9.1–11.3 mm, pupa: 3.45–3.6 mm (Hickman 1930).

Peltodytes (Neopeltodytes) edentulus (LeConte)

Plate 67.
Matheson (1912) described the egg, 3rd instar larva and the pupa.
Wilson (1924) described and illustrated the 3rd instar larva and the pupa.
Hickman (1930) described the egg, the 1st (as not discernable from *P. lengi*, which is synonymised with *P. duodecimpunctatus*), 2nd and 3rd instar larva and the pupa.
Hickman (1931) observed the copulation, and also the feeding habits of the larvae.
Measurements: total length: egg: 0,4–0,5 mm, 1st instar: 0.65–0.8 mm, 2nd instar: 2.8–4.2 mm, 3rd instar: 6.0–11.3 mm, pupa: 3.45–3.6 mm (Matheson 1912, Wilson 1924, Hickman 1930).

Peltodytes (Neopeltodytes) litoralis Matheson

Plate 68.
Orrell (2021) presented a drawing of the 3rd instar larva by Anne E. Lacy.
Measurements: total length, 3rd instar: 7.2 mm (excl. tubercles) (Orrell 2021).

Peltodytes (Neopeltodytes) muticus (LeConte)

Plate 69. Figs. 1, 2, 4, 5.
Matheson (1912) described the egg, the 1st, 2nd and 3rd instar larva and the pupa, and presented photographs of the eggs (together with eggs of *P. edentulus*) and of the larva, the larva in the pupal chamber, and the pupa.
Measurements: total length, egg: 0.466 mm, 1st instar: 2 mm, 2nd instar: 3.6 mm, 3rd instar: 6 mm, pupa: 3.5 mm (Matheson 1912).

Peltodytes (Neopeltodytes) sexmaculatus Roberts

Plate 69. Fig. 3.
Hickman (1930) described and illustrated the egg, the 1st, 2nd and 3rd instar larva and the pupa.
Hickman (1931) observed the feeding habits of the larvae.
Measurements: total length, egg: 0.4–0.5 mm, 1st instar: 0.65–0.8 mm, 2nd instar: 2.8–4.2 mm, 3rd instar: 9.1–11.3 mm, pupa: 3.45–3.6 mm (Hickman 1930).

Neotropical Region

Haliplus (Haliplidius) subseriatus Zimmermann

Plate 70. Figs. 1–16.
Vondel (2001) described and illustrated the 3rd instar larva.
Archangelsky & Michat (2014) presented some figures of the 3rd instar larva.
Michat et al. (2020) described the chaetotaxy of the 1st, 2nd and 3rd instar larva.
Measurements: total length, 3rd instar: 8.2–10.8 mm (Vondel 2001).

Haliplus (Liaphlus) brasiliensis Zimmermann

Plate 72. Figs. 10–12.
Described and illustrated by Bertrand (1934, 1972). The larva was collected in Venezuela together with two adults of *H. brasiliensis*, an identification done by M.R. Peschet as "probably this species". *H. brasiliensis* is very rarely found and only known from the holotype (Brasil). No further specimens are known. It is more likely that the larva belongs to another species. In Venezuela six species of *Haliplus* are known.
Measurements: total length, 3rd instar: 11.5 mm (Bertrand 1934).

Haliplus (Liaphlus) indistinctus Zimmermann

Plate 71.
Michat et al. (2020) described and illustrated the 1st and only described the 2nd instar larva.
Measurements: total length: 1st instar: 3.6 mm, 2nd instar: ? mm (Michat et al. 2020).

Haliplus (Liaphlus) valdiviensis Moroni

Plate 72. Figs. 1–9.
Moroni (1989) presented an illustrated description of the 3rd instar larva.
Measurements: total length, 3rd instar: 11.7–12.0 mm (Moroni 1989).

Afrotropical Region

Haliplus (Liaphlus) spp.

Plate 73. Figs. 1, 2.
Bertrand (1951, 1963, 1966b) reported 3rd instar larvae from Kenya (species unknown) and South Africa (probably *H. exsecratus*). Both larvae are distinctly different.

Haliplus (Liaphlus) exsecratus Guignot

Plate 73. Figs. 3, 4.
Bertrand (1963, 1966a, 1969) reported a 3rd instar larva from Kenya, probably belonging to this species. The head and first two thoracic segments and the last three abdominal tergites are figured. He also reported a 3rd instar larva of a *Haliplus* from South Africa, showing the same figures, but without suggesting which species this might concern.
Bertrand (1966a) reported the collection of a larva of *Haliplus* sp. (probably *H. exsecratus*) from South Africa.

Peltodytes (Peltodytes) quadratus Régimbart

Plate 73. Figs. 5.
Bertrand (1963) reported the collection of a 3rd instar larva and presented a picture of the basal tubercle on the filamentous tracheal gills. The larvae were (partly?) raised to pupa and adult, confirming the identity (Bertrand 1968).
Bertrand & Legros (1971) reported adults reared from the larva mentioned by Bertrand (1968).

Oriental Region

Haliplus (Liaphlus) angustifrons Régimbart

Plate 37. Fig. 2.
Yamasaki (2024) observed larvae living on *Chara braunii* C.C. Gmelin and on *Chara coralline* var. *kyusyensis* Imahori. He presented a photo of the 3rd instar. It is about 10.0 mm long.

Australasian Region

Haliplus (Liaphlus) halsei Watts & McRae

Plate 74, 75.
Vondel (2012) described and illustrated the 1st and 3rd instar larva.
Measurements: total length, 1st instar: 3.4 mm, third instar: 10.9–15.6 mm (Vondel 2012).

Haliplus (Liaphlus) pilbaraensis Watts & McRae

Plate 76.
Vondel (2012) described and illustrated the 3rd instar larva.
Measurements: total length: 3rd instar: 9.4–12.5 mm (Vondel 2012).

Haliplus (Liaphlus) testudo Clark

Plate 77.
Vondel (2004) described and illustrated the 3rd instar larva.
Measurements: total length: 3rd instar: 12.9 mm (Vondel 2004).

Haliplus (Liaphlus) timmsi Vondel

Plate 78.
Vondel (2004) described and illustrated the 3rd instar larva.
Measurements: total length: 3rd instar: 9.8–12.0 mm (Vondel 2004).

Phalilus fortescueensis Watts & McRae

Plate 79.
Vondel (2012) described and illustrated the 3rd instar larvae.
Measurements: total length: 3rd instar: 8.0-10.4 mm (Vondel 2012).

Phalilus pinderi Watts & McRae

Plate 80.
Vondel (2012) described and illustrated the 3rd instar larvae.
Measurements: total length, 3rd instar: 7.8–9.7 mm (Vondel 2012).

New and additional figures and descriptions

A number of results from the author's rearing project in 1986/1987 remained unpublished. Material from that project together with some larvae obtained via other sources is treated here. All material is in the author's collection.

Haliplus (Apteraliplus) parvulus

Material examined: 2 ex. 3rd instar larvae: USA, California, Dan Mateo County, Stanford Campus, Lake Lagunitas, leg. C.Y. Kitayama, 4, iv,1983.
In addition to the description of the 3rd instar (Leech & Chandler 1956) figures of the body, mandible, antenna and foreleg are presented here. Total length 3rd instar 5.3–5.9 mm. Plate 59.

Brychius elevatus

Material examined: 2 ex. 2nd + 1 ex. 3rd instar larva:

France, Clervaux les Lacs, 10.viii.1991, leg. B. J. van Vondel. Length 2nd instar: 4.0 -5.1 mm, 3rd instar. 1 ex. 3rd instar larva: France, Pyr. Atl., Bruges, 27.vii.1990, leg. B. J. van Vondel Length 9.7 mm. 4 ex. 3rd instar larva: France, St. Martin S. Nohain, 24.vii.1985, leg. B. J. van Vondel. Length 9.4-10.5 mm. 1 ex. 2nd + 1 ex. 3rd instar larva: France, Jura, Baresia sur l'Ain, 10.viii.1991, leg. B. J. van Vondel. Length 2nd instar 5.6 mm, 3rd instar 10.2 mm.

The dorsal view of the body, the legs, the mandible and the antenna of the 2nd instar larva are illustrated here as an addition to the description of Bertrand (1928b). Plate 4. Figs. 4–11.

Haliplus (Haliplidius) confinis

Material examined: 1 ex. 3rd instar larva: Netherlands, Oud Alblas, ix.1972, leg. B. J. van Vondel. Length 8,1 mm. 1 ex. 1st instar larva: Reared from adults collected in The Netherlands, Oud Alblas, 8–21.v.1986, leg. B. J. van Vondel. Length 1st instar 2.5 mm. 8 ex. 3rd instar larva: The Netherlands, Oud Alblas, 29.ix.1985, leg. B. J. van Vondel. Length 3rd instar 8.0–9.0 mm. 1 ex. 2nd + 2 ex. 3rd instar larva: The Netherlands, Naardermeer, 14.vii.1923, ex. coll. ITZ Amsterdam. Length 2nd instar 5.0 mm, 3rd instar 7,5–8,4 mm.

Description of the 1st instar larva (Plate 8. Figs. 1–10).

Body slender, widest at the thorax. Colour pale yellow. Total length 2.5 mm. Distal part (35–40%) of abdominal segment X forked. Thoracic and abdominal segments with long tubercles on margins and on dorsal side, which are set with a few denticles. Mandible with tooth on inner margin. Forelegs lacking grasping device.

Description of the 2nd instar larva (Plate 9. Figs. 1–9).

Body slender, slightly narrowing towards apex. Colour pale yellow to yellow-brown. Total length 5.0 mm. Distal part (25%) of abdominal segment X forked. Thoracic and abdominal segments with marginal and dorsal tubercles. Mandible with tooth on inner margin. Forelegs lacking grasping device.

Additional notes on the 3rd instar larva (Plate 10, 11). Total length 7.5–9.0 mm.

Haliplus (Haliplidius) obliquus

Material examined: 1 ex. 2nd instar larva: Reared from adults collected in The Netherlands, Velthunten, Kleiputten, 18.iv.1987, leg. B. J. van Vondel.

Description of the 2nd instar larva (Plate 12. Figs. 1–8).

Body slender, slightly narrowing towards apex. Colour pale yellow to yellow-brown. Total length 5.2 mm. Distal part (15%) of abdominal segment X forked. Thoracic and abdominal segments with marginal/apical extensions, set with small tubercles. Pronotum dorsally with about fourteen tubercles, mesonotum, metanotum and abdominal segments I–VII dorsally each with four tubercles, segments VIII and IX dorsally with only two tubercles. Mandible with sharp tooth on inner margin. Forelegs lacking grasping device.

Haliplus (Haliplidius) varius

Material examined: 1 ex. 2nd instar larva: Reared from adults collected in The Netherlands, Het Hol, Kortenhoef, 11.iv.1987, leg. B. J. van Vondel.

Description of the 2nd instar larva (Plate 15. Figs. 1–7).

In addition to the description of Vondel (1996) figures are here presented for the dorsal side of the body, mandible, antenna and legs.

Note: Vondel (1996) discussed the subgeneral status of this species, based on the presence of a grasping device on the forelegs.

Haliplus (Haliplus) fluviatilis

Material examined: 2 ex. 2nd + 3 ex. 3rd instar larva: Reared from adults collected: The Netherlands, Gorinchem, 2.v.1987, leg. B. J. van Vondel.

Description of the 2nd instar larva (Plate 18. Figs. 1–9).

Body slender, slightly narrowing to apex. Colour pale yellow to yellow-brown. Total length 5.2–5.5 mm. Distal part (12%) of abdominal segment X forked. Thoracic and abdominal segments laterally with small tubercles. Pronotum dorsally with about twenty-four tubercles, mesonotum, metanotum and abdominal segments I-VII dorsally with about ten tubercles, segments VIII and IX dorsally with nine tubercles. Mandible with sharp tooth on inner margin. Forelegs with thumb-like tibial projection as grasping device.

Description of the 3rd instar larva (Plate 19. Figs. 1–10).

In addition to the description by Bertrand (1950) figures are presented here for the dorsal view of the body, the antenna and the legs. Forelegs with thumb-like tibial projection as grasping device. Total length 6.8–8.5 mm.

Haliplus (Haliplus) fulvicollis

Material examined: 1 ex. 3rd instar larva: collected: The Netherlands, Kampina, 6.v.1989, leg. B. J. van Vondel and 1 ex. 3rd instar larva reared from adults from same sample. Length 6.8–7.8 mm. 1 ex. 3rd instar larva: Denmark, Vindinge, SE of Roskilde, vi.1982, leg. M. Holmen. Length 6.2 mm.
Description of the 3rd instar larva (Plate 20. Figs. 1–13).
Body slender, slightly narrowing towards apex. Colour yellow to yellow-brown. Total length 6.2–7.8 mm. Distal part (28%) of abdominal segment X forked. Margin of thoracic and abdominal segments with small tubercles and with strong apical extension. Pronotum dorsally with a number of tubercles. Thoracic segments and abdominal segments I–VII dorsally each with two rows of tubercles, ending in 3 or 4 closely set tubercles; abdominal segment VIII with four rows of tubercles, two of them ending in 3 or 4 closely set tubercles; segment IX with two dorsal rows of tubercles. Mandible with sharp tooth on inner margin. Forelegs with thumb-like tibial projection as grasping device.

Haliplus (Haliplus) furcatus

Material examined: 4 ex. 3rd instar larva: The Netherlands, Renesse, 10.viii.1985, leg. B. J. van Vondel.
Description of the 3rd instar larva (Plate 21. Figs. 1–8).
Body slender, slightly narrowing towards apex. Colour yellow to yellow-brown. Total length 6.3–6.6 mm. Distal part (20%) of abdominal segment X forked. Margin of thoracic and abdominal segments with small tubercles and with distinct apical extension. Pronotum dorsally with more than 100 tubercles, posterior edge with four groups of closely set tubercles. Mesonotum, metanotum and abdominal segments I–VII dorsally with irregular rows of tubercles, ending in 3 or 4 closely set tubercles; abdominal segment VIII with irregular rows of tubercles ending in 3 or 4 closely set tubercles; segment IX with two irregular rows of tubercles. Mandible with sharp tooth on inner margin. Forelegs with thumb-like tibial projection as grasping device.

Haliplus (Haliplus) heydeni

Material examined: 4 ex. 1st instar, 4 ex. 2nd instar and 2 ex. 3rd instar larva: Reared from adults collected in The Netherlands, Zetten, 18.iv.1987, leg. B. J. van Vondel. Length 1st instar 2.0–2.2 mm, 2nd instar 4.1–4.8 mm, 3rd instar 6.6–7.8 mm.
Description of the 1st instar larva (Plate 22. Figs. 1–6).
Body slender, slightly narrowing towards apex, colour pale yellow. Total length 2.0–2.2 mm. Distal part (60%) of abdominal segment X forked. Head with four tubercles dorsally and two laterally behind each eye. Pronotum laterally with four tubercles on each side and dorsally with ten tubercles. Meso- and metanotum with two tubercles on each lateral side and six tubercles on dorsal side. Abdominal segments with one strong tubercle laterally and one dorsally, with six tubercles on each of segments I–VII, four tubercles on segment VIII and two on segment IX. Mandible with sharp tooth on inner margin. Forelegs with thumb-like tibial projection as grasping device.
Description of the 2nd instar larva (Plate 23. Figs. 1–11).
Body slender, slightly narrowing towards apex, colour pale yellow to yellow-brown. Total length 4.1–4.8 mm. Distal part (3%) of abdominal segment X forked. Head with a few dorsal setae and laterally two tubercles behind each eye. Margins of thoracic and abdominal segments set with small tubercles. Pronotum dorsally with about 26 tubercles. Meso- and metanotum and abdominal segments I–VII each dorsally with twelve tubercles, segment VIII with ten tubercles and segment IX with eight tubercles on dorsal side. Mandible with sharp tooth on inner margin. Forelegs with thumb-like tibial projection as grasping device.
Description of the 3rd instar larva (Plate 24. Figs. 1–15).
In addition to the description by Bertrand (1942) and Vondel (1997) figures are added here for the body in dorsal view, the antenna and the legs. Forelegs with thumb-like tibial projection as grasping device. Length 6.6–7.8 mm.

Haliplus (Haliplus) immaculatus

Material examined: 3 ex. 1st instar larva: reared from adults collected in The Netherlands, Renesse, Oude Hoeve, 3.v.1986, leg. B. J. van Vondel. Length 1.5–2.5 mm. 2 ex. 1st instar larva: reared from adults collected in: The Netherlands, Zetten, 10.v.1986, leg. B. J. van Vondel. Length 1.3 mm. 1 ex. 2nd instar larva: reared from adults collected in The Netherlands, Renesse, Oude Hoeve, 3.v.1986, leg. B. J. van Vondel. Length 3.8 mm. 1ex. 3rd instar larva: reared from

adults collected in The Netherlands, Doetinchem, 14.vi.1986, leg. B. J. van Vondel. Length 5.9 mm.
Description of the 1st instar larva (Plate 25. Figs. 1–13).
In addition to the illustrated description by Falkenström (1926) figures are here added for the larva in dorsal and lateral view, the antenna, the mandible and the legs. Forelegs with thumb-like tibial projection as grasping device. Total length 1.3-2.5 mm.

Description of the 2nd instar larva (Plate 26. Figs. 1–11).
In addition to the description of Falkenström (1926) without figures, illustrations are added here for the larva in dorsal and lateral view, the antenna, the mandible and the legs. Forelegs with thumb-like tibial projection as grasping device. Total length 3.8 mm.
Description of the 3rd instar larva (Plate 27. Figs. 1–15).
In addition to the descriptions of Falkenström (1926), Bertrand (1950) and Vondel (1997) a figure of the body in dorsal view is added here. Total length 5.9 mm.

Haliplus (Haliplus) lineolatus

Material examined: 8 ex. 1st +8 ex. 2nd + 12 ex. 3rd instar larva: Reared from adults collected in The Netherlands, Zetten, 10.v.1986. Length 1st instar 1.7–2.5 mm, 2nd instar 3.8–5.1 mm, 3rd instar 6.4–8.2 mm.
Description of the 1st instar larva (Plate 28. Figs. 1–11).
Body slender, slightly narrowing towards apex, colour pale yellow. Total length 1.7–2.5 mm. Distal part (60%) of abdominal segment X forked Head with six strong tubercles dorsally and two laterally behind each eye. Pronotum laterally with four tubercles on each side and dorsally with ten tubercles. Meso- and metanotum with two tubercles on each lateral side and six tubercles on dorsal side. Abdominal segments each with one strong tubercle laterally, and dorsally with six tubercles on segments I–VII, four tubercles on segment VIII and IX. Mandible with sharp tooth on inner margin. Forelegs with thumb-like tibial projection as grasping device.
Description of the 2nd instar larva (Plate 29. Figs. 1–12).
Body slender, slightly narrowing towards apex, colour pale yellow to yellow-brown. Total length 3.8–5.1 mm. Distal part (25%) of abdominal segment X forked. Head with a few dorsal setae and laterally two tubercles behind each eye. Margins of thoracic and abdominal segments set with small tubercles. Pronotum dorsally with about 24 tubercles. Meso- and metanotum and abdominal segments I-VII dorsally with twelve tubercles, abdominal segments VIII and IX each with ten tubercles on dorsal side. Mandible with sharp tooth on inner margin. Forelegs with thumb-like tibial projection as grasping device.
Description of the 3rd instar larva (Plate 30. Figs. 1–15).
In addition to the descriptions of Bertrand (1931), Seeger (1971) and Vondel (1997) figures are added here for the body in dorsal view, the antenna and the legs. Forelegs with thumb-like tibial projection as grasping device. Length 6.4–8.2 mm.

Haliplus (Haliplus) ruficollis

Material examined: 2 ex. 1st instar + 2 ex. 2nd instar + 2 ex. 3rd instar larva, reared from adults collected in The Netherlands, Renesse, 3.v.1986. 1 ex. 2nd instar, 2 ex. 3rd instar larva reared from adults collected in The Netherlands, Oud Alblas, 8.v.1986, leg. B. J. van Vondel.
Description of the 1st instar larva (Plate 31. Figs. 1–9).
In addition to the description of Bertrand (1923) figures are added here for the body in dorsal and lateral view, the antenna, the mandible and the legs. Forelegs with thumb-like tibial projection as grasping device. Length 1.8–1.9 mm.
Description of the 2nd instar larva (Plate 32. Figs. 1–9).
In addition to the descriptions of Bertrand (1923, 1928b) figures are added here for the body in dorsal view, the antenna, the mandible and the legs. Forelegs with thumb-like tibial projection as grasping device. Length 3.6–4.9 mm.
Description of the 3rd instar larva (Plate 33. Figs. 1–15).
In addition to the descriptions of Schiødte (1864) and Bertrand (1923, 1928b, 1931) figures are added here for the body in dorsal view and the legs. Forelegs with thumb-like tibial projection as grasping device. Length 6.8–9.5 mm.

Haliplus (Haliplus) sibiricus

Material examined: 9 ex. 1st instar larva, 5 ex. 2nd instar larva, 2 ex. 3rd instar larva: Reared from adults collected in The Netherlands, Zetten, 10.v.1986.

Description of the 1st instar larva (Plate 34. Figs. 1–7).
Body slender, slightly narrowing towards apex, colour pale yellow. Total length 1.8–2.5 mm. Distal part (50%) of abdominal segment X forked. Head with about four strongly developed tubercles dorsally and two laterally behind each eye. Pronotum laterally with three tubercles on each side and dorsally with six tubercles. Meso- and metanotum with two tubercles on each lateral side and four tubercles on dorsal side. Abdominal segments each with one strong tubercle laterally, and dorsally with four tubercles. Mandible with sharp tooth on inner margin. Forelegs with thumb-like tibial projection as grasping device. Beier (1929) figured the head, thorax and first two abdominal segments, but probably he actually described the larvae of *Haliplus lineatocollis*. See discussion earlier in this study.
Description of the 2nd instar larva (Plate 35. Figs. 1–6).
Body slender, slightly narrowing towards apex, colour pale yellow to yellow-brown. Total length 3.7–5.3 mm. Distal part (25%) of abdominal segment X forked. Head with a few dorsal setae and laterally two tubercles behind each eye. Margins of thoracic and abdominal segments set with small tubercles. Pronotum dorsally with about 16 tubercles. Meso- and metanotum and abdominal segments I–VII each dorsally with about 14 tubercles, segments VIII and IX with about 12 tubercles on dorsal side. Mandible with sharp tooth on inner margin. Forelegs with thumb-like tibial projection as grasping device.
Description of the 3rd instar larva (Plate 36. Figs. 1–14).
Body slender, slightly narrowing towards apex, colour yellow to yellow-brown. Total length 6.6–7.2 mm. Distal part (20%) of abdominal segment X forked. Margin of thoracic and abdominal segments with small tubercles and with weak apical extension. Pronotum dorsally with more than 100 tubercles, posterior edge with four groups of closely set tubercles. Mesonotum, metanotum and abdominal segments I–VII each dorsally with irregular rows of tubercles, ending in 3 or 4 closely set tubercles, abdominal segment VIII with two irregular rows of tubercles ending in 3 or 4 closely set tubercles, while segment IX has two irregular rows of tubercles ending in 3 or 4 closely set tubercles. Mandible with sharp tooth on inner margin. Forelegs with thumb-like tibial projection as grasping device.
Beier (1929) figured the head, thorax and first two abdominal segments, the mandible, fore- and midleg, but probably he actually described the larvae of *Haliplus lineatocollis*. See discussion earlier in this study.

Haliplus (Liaphlus) flavicollis

Material examined: 2 ex. 2nd instar + ca 50 ex. 3rd instar larva, The Netherlands, Oud Alblas, 29.ix.1985-1.vi.1986, leg. B. J. van Vondel. 7 ex. 3rd instar larva The Netherlands, Lage Zwaluwe, 12.ix.1993, leg. B. J. van Vondel.
Description of the second instar larva (Plate 41. Figs. 1–6).
Body wide, narrowing towards apex, colour pale yellow to yellow-brown. Total length 3.5–4.8 mm. Distal part (18%) of abdominal segment X forked. Head with a few dorsal setae and laterally one tubercle behind each eye. Margins of thoracic and abdominal segments with a strong extension set with small tubercles. Pronotum dorsally with about 6 small tubercles and two strong extensions set with small tubercles. Meso- and metanotum and abdominal segments I–IX each dorsally with two strong extensions set with small tubercles. Mandible with sharp tooth on inner margin. Forelegs lacking grasping device.
Description of the 3rd instar larva (Plate 42. Figs. 1–8).
In addition to the description of Bertrand (1928b, 1931) and Vondel (1997) pictures are added here for the larva in dorsal, lateral and ventral view and the antenna. Length 7.5–12.0 mm.

Discussion

Gaps of knowledge

Of about 25% of the known species of Haliplidae worldwide, one or more immature stages were collected, reported and described in various ways, ranging from a single structural detail to extensively illustrated studies with or without photographs. Most is known from the European species, followed by species from the Nearctic, Australian and Japanese regions. Less is known from the Neotropical, Afrotropical and Oriental fauna. It is advised to apply the specific collecting methods necessary for finding haliplid larvae (Seeger 1971a, Vondel 1997). Larvae are slender and not very active, so they are often overlooked. In the field a good method to collect them is to gather the algae (or other aquatic plants) they usually live on and put them on a wide-meshed sieve above a white tray with some water. After some time, even hours or days, the larvae leave the drying plants and fall into the water. Rearing is a good method to gain more knowledge about the different immature stages. It should be encouraged that collecting adults for research projects will be accompanied by search for larvae.

Phylogeny

The phylogenetic relationships of Haliplidae with other groups of Adephaga was studied several times, but was usually based on adults, with the exception of Beutel (1993). The phylogeny of the genera, subgenera and species of Haliplidae was studied by Beutel & Ruhnau (1990) and Vondel (2019), both based on adults. There are some keys to 3rd instar larvae of Haliplidae, but not within a phylogenetic context.

Five species of *Haliplus* belong to the subgenus *Haliplidius*. Vondel (1996, 2001) described the larvae of *Haliplus* (*Haliplidius*) *varius* and *Haliplus* (*Haliplidius*) *subseriatus*, but noted that the forelegs have a grasping device on the tibia (*H. varius*) or on the femur (*H. subseriatus*), while the larvae of *Haliplus* (*Haliplidius*) *confinis* and *Haliplus* (*Haliplidius*) *obliquus* lack such a device. The larvae of *Haliplus* (*Haliplidius*) *rejseki* Šťastný & Boukal are unknown. Further research is necessary about the subgeneric status of those five species.

Michat et al (2020) described the first instar larva of *Haliplus* (*Liaphlus*) *indistinctus*, but the presence of a grasping device on the tibia and femur of the forelegs suggest that it does not belong to the subgenus *Liaphlus*.

A future study on the phylogeny of Haliplidae based on larvae is needed and should consider at least the following characters:

Characters relevant for comparison with other groups of Adephaga:

- Shape of cardo.
- Maxillary musculature.
- Sternites and pleurites of postcephalic body.
- Pygopod-like tenth abdominal segment absent.

Characters Haliplidae (automorphy):

- Mandibles with very distinctive sucking channel, suited for sucking cell contents of algae and other plants.
- Legs with a single claw.

Characters of genera, subgenera or species:

- Deposition of the eggs: on or in algae or plants.
- Feeding on Algae, Charophyta or Angiospermophyta.
- Forelegs with or without a grasping device on tibia or femur in different forms.
- Number of abdominal segments, nine in *Peltodytes*, 10 in *Brychius*, *Haliplus* and *Phalilus*.
- Tubercles posterior to the stemmata present or absent.
- Antenna with a third segment distinctly longer than two basal segments or third segment not longer than basal segments.
- Abdominal segment X (Haliplidae excl. Peltodytes) long and forked apically. Length of segment and forked part may vary considerably.
- Dorsal and lateral tracheal gills short (tubercles) or very long (segmented or not).
- Posterior edge of thoracic and abdominal segments with single tubercles or with some closely set tubercles.
- Latero-posterior corners of thoracic and abdominal segments with small or strongly developed processes or without such process (sometimes with a single seta).

Ackowledgements

The Uyttenboogaart-Eliasen Foundation is greatly acknowledged for the financial support of this project and my research in general. I am very grateful to Dr. Rolf Beutel for his valuable cooperation. Mogens Holmen (†) is thanked for sending some larvae. I am also grateful to The Institute for Taxonomic Zoology (ITZ, now incorporated in Naturalis, Leiden) for providing some larvae. Dr. N. Hayashi helped to locate Japanese studies, which is also gratefully acknowledged. Some authors (see below under credits) sent me the original photos, for which I am very thankful.

Credits

To prevent problems with copyright I have contacted the authors/publishers of which I reproduced drawings or photographs published in the last 60 years. The given permissions for which I am very

grateful, are listed here.
The Schweizerbart'sche Verlagsbuchhandlung OHG (Dr. Andreas Nägele) gave permission to reuse the figures of Seeger (1971a).
Dr. Rasmus Keis Neerbeek of the Entomologisk Forening gave permission to use the figures of Holmen (1981).
Dr. Anders Nilsson gave permission to use some figures of Nilsson (1982).
Dr. Jinze Noordijk of the Nederlandse Entomologische Vereniging told me that the copyright of Vondel (1986,1996, 2001) belongs to the author.
Brill publishers (Dr. Tanja Cowall) gave permission to reuse figures of Holmen (1987) and Vondel (2004, 2011b, 2012).
Dr. Jerry L. Cook gave permission (as editor of the Proceedings of the Entomological Society of Washington) to use the pictures of *Brychius hungerfordi* in Strand & Spangler (1994).
Dr. Tonya Mousseau gave me permission to reuse the figures from her thesis (2004).
Dr. Donald Chandler sent me the original photo of *Peltodytes duodecimpunctatus* (Chandler 2008).
Dr. Vasily Dyadichko gave me permission to use the photo of the larva of *Haliplus zacharenkoi* (Dyaditchko (2009).
Dr. Masakazu Hayashi sent me the original photo of *Haliplus ovalis* (Hayashi et al. 2011).
Dr. Manfred Jäch (as editor of Koleopterologische Rundschau) gave permission to use the figures from Vondel (2011a).
Dr. Kosuke Nakanishi sent me the original photos of *Haliplus sharpi* (Nakanishi et al. 2012).
Dr. Rolf Beutel gave permission (as editor of Handbook of Zoology) for the use of photos of *Brychius elevatus* (Vondel 2016) and as author for the figure of *Brychius elevatus* (Beutel 1997).
Dr. Mariano Michat gave me permission to use the figures of *Haliplus subseriatus* and *H. indistinctus* (Archangelsky & Michat 2014, Michat & Archangelsky 2020).
Via Dr. Toshimasa Mitamura the publisher gave permission to use the photos of *Haliplus basinotatus*, *H. japonicus*, *H. sharpi* and *Peltodytes intermedius* (Mitamura et al. 2017).
Dr. Hiroyuki Yoshitomi gave permission to use the figures of *Peltodytes intermedius* and *Haliplus ovalis* in Sato & Yoshitomi (2018).
Dr. Masakazu Hayashi gave permission to reuse the photo of *Haliplus japonicus* and Dr. Tadashi Kitano sent me photos of eggs, third instar larva and pupa of *Peltodytes intermedius*, from the same material as presented in (Nakajima et al. 2020).
Dr. Reiya Watanabe sent the original photos of *Haliplus kamiyai* (Watanabe 2020).
Dr. Shun Yamasaki gave permission to use the photos of *Haliplus eximius* and *H. angustifrons* (Yamasaki et al. 2020, Yamasaki 2024).
Dr. Masakazu Hayashi sent me the original photo of *Peltodytes intermedius* (pers. comm.).
Dr Fenglong Jia and Dr. Zuqi Mai gave permission to use the photo of *Peltodytes sinensis* (Jia et all. 2024).

For the following references I rely on a Creative Commons License (or a comparable arrangement):
Steinmann (1984): The drawing of the antenna of the larva of *Peltodytes caesus* is wrong, but included here to prevent misunderstanding. I assume that giving the reference is sufficient.
Beutel (1986): a proper reference is obliged and given here.
Moroni (1989): a proper reference is obliged and given here.
Wichard (1995): the publisher Gustav Fischer stopped in 2008 and the author deceased in 2024. I assume that giving the reference is sufficient.
Vondel (1997): the publisher Gustav Fischer stopped in 2008. I assume that the copyright returns to the author.
Orrell (2012): pictures of *Peltodytes litoralis*. Usage permitted with reference (Usage Conditions apply).

References

Archangelsky, M., & Michat, C. (2014). Haliplidae. In S. Roig–Juñent, L. E. Claps, & J. J. Morrone (Eds.), *Biodiversidad de Artrópodos Argentinos*.

Beier, M. (1929). Zur Kenntnis der Lebenweise von *Haliplus wehnckei* Gerh. *Zeitschrift für Morphologie und Ökologie der Tiere, 14*(1), 191–233. https://doi.org/10.1007/BF00419347

Bertrand, H. (1923). Larve et Nymphe: d'Haliplus lineaticollis Marsh [Coléopt.]. *Annales de la Société Entomologique de France, 92*(4), 323–338. https://doi.org/10.1080/21686351.1923.12279951

Bertrand, H. (1925a). Liste des larves des Dytiscides, Hygrobiides et Haliplides actuellement connues; observations sur diverses espèces Françaises. *Encyclopédie entomologique. 1. Coleoptera, 1*, 1–23.

Bertrand, H. (1925b). Captures et élevages de larves de Coléoptères aquatiques. *Annales de la Société Entomologique de France, 94*(4), 441–446. https://doi.org/10.1080/21686351.1925.12280177

Bertrand, H. (1928a). Captures et élevages de larves de

coléoptères aquatiques (3e note). *Annales de la Société Entomologique de France, 97*(3–4), 419–424. https://doi.org/10.1080/21686351.1928.12280193

Bertrand, H. (1928b). Les larves et nymphes des Dytiscides, Hygrobiides et Haliplides. *Encyclopedié Entomologique 10*, I–VI+366 p. Paris.

Bertrand, H. (1928c). Les larves et nymphes des Dytiscides, Hygrobiides et Haliplides. *Encyclopedié Entomologique 10*, I–VI+366 p. Paris.

Bertrand, H. (1930). Captures et élevages de larves de coléoptères aquatiques (4e note). *Annales de la Société Entomologique de France, 99*(1), 65–77. https://doi.org/10.1080/21686351.1930.12279987

Bertrand, H. (1931). Captures et élevages de larves de coléoptères aquatiques (5e note). *Annales de la Société Entomologique de France, 100*(1-2), 87–98. https://doi.org/10.1080/21686351.1931.12280123

Bertrand, H. (1933). Les larves aquatiques de Coléoptères. II–Les Ampnizoïdes, III–Les Hygrobiides, IV–Les Haliplides, V–Les Gyrinides. *La Terre et la Vie, 3*(9), 523–534. https://doi.org/10.3406/revec.1933.2971

Bertrand, H. (1934). Notes sur quelques larves de Coléoptères aquatiques. *Annales de la Société Entomologique de France, 103*(3–4), 363–382. https://doi.org/10.1080/21686351.1934.12280244

Bertrand, H. (1942). Captures et élevages de larves de coléoptères aquatiques (9e note). *Annales de la Société Entomologique de France, 111*(1), 67–74. https://doi.org/10.1080/21686351.1942.12279001

Bertrand, H. (1949). Notes biologiques sur les larves des Haliplides (Col.). *Bulletin de la Société Entomologique de France, 54*(6), 91–95. https://doi.org/10.3406/bsef.1949.18380

Bertrand, H. (1950). Captures et élevages de larves de coléoptères aquatiques (11e note). *Bulletin de la Société Entomologique de France, 55*(3), 42–48. https://doi.org/10.3406/bsef.1950.18429

Bertrand, H. (1951). Captures et élevages de larves de coléoptères aquatiques (13e note). *Bulletin de la Société Entomologique de France, 56*(5), 75–80. https://doi.org/10.3406/bsef.1951.18507

Bertrand, H. (1954). *Les Insectes Aquatiques d'Europe (Genres: Larves, Nymphes, Imagos).* Paris: Paul Lechevalier.

Bertrand, H. (1963). Contribution à l'étude des premiers états des Coléoptères aquatiques de la region éthiopienne. (5e note). Familles: Haliplidae, Dytiscidae, Gyrinidae (Hydrocanthares). *Bulletin de l'Institut Français d'Afrique Noire, 25*(2), 390–402.

Bertrand, H. (1966a). Récoltes de larves de Coléoptères aquatiques dans la région éthiopienne: Afrique du Sud et Rhodesies (1959). *Bulletin de la Société Entomologique de France, 71*(9), 252–261. https://doi.org/10.3406/bsef.1966.20900

Bertrand, H. (1966b). Larves de Coléoptères aquatiques de l'Afrique Orientale. *Bulletin du Muséum National d'Histoire Naturelle 2e serie, Tome 38, No. 5*, pp 562–573.

Bertrand, H. (1968). Récoltes de larves de ColéoptèresAquatiques dans la Région Éthiopienne: Madagascar (1960). *Bulletin du Muséum National d'Histoire Naturelle 2e Série, Tome 40*(4), 718–733.

Bertrand, H. (1969). Larves de Coléoptères aquatiques récoltées par M. et Mme J. Omer–Cooper en Afrique du Sud (suite et fin). *Bulletin de la Société Entomologique de France, 74*(7), 156–162. https://doi.org/10.3406/bsef.1969.21075

Bertrand, H. (1972). *Larves et Nymphes des Coléoptères Aquatiques du Globe. Avec tableaux de determination des Genres.* Abbeville: F. Paillart.

Bertrand, H. & Legros, C. (1971). Hydrocanthares (excl. Gyrinidae) receuillis à Madagascar. Mission Bertrand, 1960). *Cah. O.R.S.T.O.M., sér. Hydrobiol., vol. V*, nom 3/4, 241–249.

Beutel, R. (1986). Skelet und Muskulatur des Kopfes der Larve von *Haliplus lineatocollis* Mrsh. (Coleoptera: Haliplidae). *Stuttgarter Beiträge zur Naturkunde* [rec. 1987]. *Ser. A., 390*, 1–15.

Beutel, R. G. (1993). Phylogenetic analysis of Adephaga (Coleoptera) based on characters of the larval head. *Systematic Entomology, 18*(2), 127–147. https://doi.org/10.1111/j.1365-3113.1993.tb00658.x

Beutel, R. G. (1997). Über Phylogenese und Evolution der Coleopteren (Insecta) insbesondere der Adephaga. *Abhandlungen des Naturwissenschaftlichen Vereins in Hamburg. (NF) 31*, 164 pp.

Beutel, R. G., & Ruhnau, S. (1990). Phylogenetic analysis of the genera of Haliplidae (Coleoptera) based on characters of the adults. *Aquatic Insects, 12*(1), 1–17. https://doi.org/10.1080/01650429009361381

Beutel, R. G., & van Vondel, B. J. (2024). Misidentification of fossil beetle larvae on the subordinal level–Scraptiidae (Polyphaga: Tenebrionoidea) instead of Haliplidae (Adephaga). *Koleopterologische Rundschau, 94*, 187–192.

Böving, A. G. & Craighead, F. C. (1930–1931). An illustrated synopsis of the principal larval forms of the order Coleoptera. *Entomologica Americana 11 (N.S.)* (1), 1–351, 125 pl.

DeRyke, W. (1922).The Food of the Fishes on Winona Lake in Kosciusko County During the Months of June, July and August. *The Department of Conservation State of Indiana, The Division of Fish and Game, Publication No. 29*, 48 pp.

Dyadichko, V. G. (2009). *Aquatic carnivorous beetles (Coleoptera, Hydradephaga) North–Western Black Sea. Odessa*, (204 pp). Publishing House "Astroprint", [in Russian].

Falkenström, G. (1926). Beiträge zur Kenntnis der Biologie der Haliplidae und der Metamorphose von *Haliplus immaculatus* Gerh. *Entomologisk Tidskrift, 47*, 1–28.

Fukuda, A., Kurosa, K., & Hayashi, N. (1959). Coleoptera. In A. Kawada (Ed.), *Illustrated insect larva of Japan* (pp. 392–545). Tokyo: Hokuryo–kan. [in Japanese]

Gernet, C von. (1868). Beiträge zur Käferlarvenkunde. *Horae Societatis Entomologica Rossicae. Petersburg, 5*, 7–22.

Hayashi, M. (2011). Aquatic Coleoptera of Shimane Prefecture. *Spec. Bull. Hoshizaki Green Found, (1)*, 1–117.

Henriksen, K. (1930). Vandkalve– og Hvirvlerlaverne. In V. Hansen (Ed.), *Biller, VIII. Vandkalve og Hvirvlere (Haliplidae, Dytiscidae & Gyrinidae).* Danmarks Fauna 34, 1–233.

Hickman, J. R., (1930). Life–histories of Michigan Haliplidae (Coleoptera). *Studies of the Michigan Academy (1929), 12*, 399–424.

Hickman, J. R. (1931). Contributions to the biology of the

Haliplidae (Coleoptera). *Annals of the Entomological Society of America, 24*(1), 129–142. https://doi.org/10.1093/aesa/24.1.129

Hickman, J. R. (1937). Family Haliplidae. In P. S. Galtsoff, F. E. Lutz, P. S. Welch, & J. G. Needham (Eds.). *Culture Methods for Invertebrate Animals*. New York.

Hinton, H. E. (1947). On the reduction of functional spiracles in the aquatic larvae of the Holometabola with notes on the moulting process of spiracles. *Transactions of the Royal Entomological Society of London, 98*(10), 449–473. https://doi.org/10.1111/j.1365-2311.1947.tb01222.x

Holmen, M. (1981). Status over Danmarks Haliplidae (Coleoptera) med bemaerkninger om zoogeografi og auto/kologi. *Entomologiske Meddelelser, 49*, 1–14.

Holmen, M. (1987). The aquatic Adephaga (Coleoptera) of Fennoscandia and Denmark. I. Gyrinidae, Haliplidae, Hygrobiidae and Noteridae. *Fauna Entomologica Scandinavica. 20*, 9–17, 1–168.

Jaboulet, H. C., (1960). Contribution à l'étude des larves d'Haliplides. *Travaux du Laboratoire de Zoologie et de la Station aquicole Grimaldi de la Faculté des Sciences de Dijon 31*,1–15, 30 figs.

Jia, F., Mai, Z., Jiang, Z., & Liang, Z. (2024). *Handbook of Aquatic Beetles in China (Myxophaga, Hydradephaga, Hydrophiloidea)*. Straits Books Publishing Home. [in Chinese]

Kitano, T., & Sano, S. (2011). Haliplidae larvae eat algae. In H. Karube (Ed.), *Suisei Kontyu daihyakka, Kanagawa prefectural Museum* (pp. 48–49). Odawara. [in Japanese]

Klausnitzer, B. (1977). Bestimmungstabellen für die Gattungen der aquatischen Coleopteren–Larven Mitteleuropas. 22. Beitrag zur Kenntnis der mitteleuropäischen Coleopteren–Larven. *Beiträge zur Entomologie, Berlin, 27*, 145–192.

Klausnitzer, B. (1984). *Käfer im und am Wasser* (Vol. 567). A. Ziemsen Verlag DDR.

Klausnitzer, B. (1991). *Die Larven der Käfer Mitteleuropas. 1. Band.Adephaga, 3e Familie: Haliplidae*. Pp. 142–154. Krefeld: Goecke & Evers.

Klausnitzer, B. (1996). *Käfer im und am Wasser. Die Neue Brehm Bücherei 567*, 200 pp. Magdeburg.

Leech, H. B., & Chandler, H. P. (1956). Aquatic Coleoptera. Chapter 13. In R. L. Usinger (Ed.) *Aquatic Insects of California. With keys to North American genera and California species*, Ix + 508 pp. Berkeley and Los Angeles, California: University of California Press

Leprieur, M. C. E. (1887). Séance du 27 avril 1887. Bulletins des Séances et Bulletin Bibliographique de la Société Entomologique de France. *Annales de la Société Entomologique de France 6e série, Vol. 7.*

Linhart, S. J., Müller, P., Haug, G. T., Haug, C., & Haug, J. T. (2023). An overview of crawling water beetle larva and a first possible record from 100–million–years–old Myanmar amber. *Palaeontologia Electronica, 26*(3), 1–20.

Makarov, K. V. & Prokin, A. (2015). About homology of *Haliplus* Latreille 1802 larvae postanal process and other structural features of Haliplidae. Powerpoint presentation, 2015 at the Immature Beetles Meeting 2015. *Acta Entomologica Musei Nationalis Pragae 55*(2), 879–881.

Matheson, R. (1912). The Haliplidae of North America, North of Mexico. *Journal of the New York Entomological Society, 20*, 156–193.

Michat, M. C., M. Archangelsky, & Alarie, Y. (2020). Morphology and chaetotaxy of Neotropical Haliplidae larvae (Coleoptera: Haliplidae). *Revista Mexicana de Biodiversidad 91*, e913541. 12 pp.

Mitamura, T., Hirasawa, K., & Yoshii, S. (2017). *The handbook of Japanese aquatic insects, volume 1, Coleoptera*, (352 pp). Tokyo: Bun–ichi Sogo Shuppan. [in Japanese].

Moroni, J. (1989). Aporte al conocimiento de los Haliplidae de Chile. II. Descripcion del tercer instar larval de *Haliplus valdiviensis* Moroni, 1980 (Coleoptera: Dytiscoidea). *Revista Chilena de Entomologia, 17*, 89–93.

Mousseau T. (2004). Taxonomy, Classification, Reconstructed Phylogeny, Biogeography, and Natural History of Nearctic Species of *Brychius* Thomson (Coleoptera: Haliplidae). xvi+144 pp., 42 figs, 5 app. Thesis.

Mousseau, T., & Roughley, R. T. (2007). Taxonomy, Classification, reconstructed Phylogeny and Biography of Nearctic Species of *Brychius* Thomson (Coleoptera: Haliplidae). *Coleopterists Bulletin, 61*(3), 351–397. https://doi.org/10.1649/0010-065X(2007)61[351:TCRPAB]2.0.CO;2

Nakajima, J., Hayashi, M., Ishida, K., Kitano, T., & Yoshitomi, H. (2020). *Aquatic Coleoptera and Hemiptera of Japan*, (352 pp.). Japan. [in Japanese].

Nakanishi, K. (2012). Biological Notes on *Haliplus sharpi* Wehncke, 1880 (Coleoptera: Haliplidae). *Elytra, Tokyo. New Series 2,* (1), 107–108.

Nilsson, A. N. (1988). A review of primary setae and pores on legs of larval Dytiscidae (Coleoptera). *Canadian Journal of Zoology, 66*(10), 2283–2294. https://doi.org/10.1139/z88-339

Orrell T, Informatics Office (2021). NMNH Extant Specimen Records. Version 1.40. National Museum of Natural History, Smithsonian Institution. Occurrence dataset. https://doi.org/10.15468/hnhrg3 accessed via GBIF.org on 2021–02–16. https://www.gbif.org/occurrence/1317392490

Pennak, R. W. (1953). *Freshwater invertebrates of the United States.* New York: Ronald Press.

Peterson, A. (1960). Larvae of Insects. *An Introduction to Nearctic Species.II. Coleoptera, Diptera, Neuroptera, Siphonaptera, Mecoptera, Trichoptera*, (I–V, 1–416 pp). Ohio: Columbus.

Rousseau, E. (1919a). Notes limnologiques. III. La larve présumée du *Brychius elevatus* Panz. (Coléoptère). *Bulletin de la Société Entomologique de Belgique. 1919*(1), 128+163–165.

Rousseau, E. (1919b). Contribution à l'étude des larves d'Haliplides d'Europe. *Annales de Biologie Lacustre, 9*, 269–278.

Rozkošný, R. (Ed.). (1980). *Klíč vodních larev hmyzu* [keys to the waterlarvae of insects]. Prague: Academia.

Rupertsberger, M. (1879). Catalog der bekannten europäischen Käfer–Larven. *Entomologische Zeitung, 40*, 211–236.

Sato, M., & Yoshitomi, H. (2018). Coleoptera. In T. Kawai, & T. Tanida (Eds.). *Aquatic Insects of Japan: Manual with keys and illustrations. Second Edition* (pp. 707–790). Hadano: Tokai University Press. [in Japanese]

Schiødte, J. C. (1864). De Metamorphosi Eleutheratorum observationes: Bidrag til Insekternes Udviklingshistorie. Pars

II, Histri, Dytisci, Gyrini (Suppl.), Staphylinini, Oxytelini.. *Naturhistorisk Tidsskrift, 3*, 131–224.

Schiødte, J. C. (1872). De Metamorphosi Eleutheratorum observationes: Bidrag til Insekternes Udviklingshistorie. Pars VI. Carabi (Suppl.), Dytisci (Suppl.), Gyrini (Suppl.), Hydrophili (Suppl.). *Naturhistorisk Tidsskrift, 8*, 165–226.

Seeger, W. (1971a). Morphologie, Bionomie und Ethologie von Halipliden, unter besonderer Berücksichtigung funktionsmorphologischer Gesichtspunkte (Haliplidae: Coleoptera). *Archiv für Hydrobiologie, 68*, 400–435.

Seeger, W. (1971b). Autökologische Laboruntersuchungen an Halipliden mit zoogeographischen Anmerkungen (Haliplidae; Coleoptera). *Archiv für Hydrobiologie, 68*, 528–547.

Seeger, W. (1971c). Die Biotopwahl bei Halipliden, zugleich ein Beitrag zum Problem der syntopischen (sympatrischen s.str.) Arten (Haliplidae; Coleoptera). *Archiv für Hydrobiologie, 69*, 175–199.

Spangler, P. J. (1991). Haliplidae. In F. W. Stehr (Ed.), *Immature Insects* (Vol. 2, pp. 311–312).

Steinmann, H., & Zombori, L. (1984). *A morphological atlas of insect larvae*. Budapest: Akadémia Kiadó.

Strand, R. M., & Spangler, P. J. (1994). The natural history, distribution, and larval description of *Brychius hungerfordi* Spangler (Coleoptera: Haliplidae). *Proceedings of the Entomological Society of Washington, 96*, 208–213.

van Vondel, B. J. (1986). Description of the second and third instar larvae of *Haliplus laminatus* (Schaller) with notes on the subgeneric status (Coleoptera: Haliplidae). *Entomologische Berichten (Amsterdam), 46*, 128–132.

van Vondel, B. J. (1996). Description of the second and third instar larva of *Haliplus varius* with notes on the subgeneric status (Coleoptera: Haliplidae). *Entomologische Berichten (Amsterdam), 56*, 9–11.

van Vondel, B. J. (1997). Haliplidae. In J. Schwoerbel & P. Zwick (Eds.). *Süsswasserfauna von Mitteleuropa 20.2* (pp. 1–95). Stuttgart: Gustav Fischer.

van Vondel, B. J. (2001). Description of the third instar larva of *Haliplus subseriatus* (Coleoptera: Haliplidae). *Entomologische Berichten (Amsterdam), 61*, 14–16.

van Vondel, B. J. (2004). First description of larvae of *Haliplus*–species from Australia (Coleoptera: Haliplidae). *Tijdschrift voor Entomologie, 147*(1), 57–61. https://doi.org/10.1163/22119434-900000140

van Vondel, B. J. (2011a). Description of *Haliplus* larvae from Lebanon (Coleoptera: Haliplidae). *Koleopterologische Rundschau, 81*, 41–54.

van Vondel, B. J. (2011b). Description of the third instar larva of *Haliplus variomaculatus* Brigham & Sanderson with notes on larvae of Nearctic Haliplidae (Coleoptera). *Tijdschrift voor Entomologie, 154*(1), 127–133. https://doi.org/10.1163/22119434-900000310

van Vondel, B. J. (2012). Description of larvae of four *Haliplus* species from Australia (Coleoptera: Haliplidae). *Tijdschrift voor Entomologie, 155*(2-3), 193–208. https://doi.org/10.1163/22119434-00002012

van Vondel, B. J. (2016). Haliplidae. In R. G. Beutel & R. A. B. Leschen (Eds.), *Handbook of Zoology. Arthropoda: Insecta. Coleoptera, Beetles. Morphology and Systematics. Archostemata, Adephaga, Myxophaga and Polyphaga Partim* (2nd ed., Vol. 1, pp. 89–98). Berlin: Walter de Gruyter.

van Vondel, B. J., & Litovkin, S. (2017). Five new synonymies in *Haliplus* subg. *Liaphlus* Guignot, based on the variability of the left paramere (Coleoptera: Haliplidae). *Koleopterologische Rundschau, 87*, 31–35.

van Vondel, B. J. (2019). Features of the metacoxal air–storage space as additional characters for reconstructing the phylogeny of Haliplidae (Coleoptera). *Tijdschrift voor Entomologie, 162*(1), 13–32. https://doi.org/10.1163/22119434-20192081

van Vondel, B. J. (2023). Revision of the Nearctic Haliplidae (Coleoptera). Corrections and additions I. *Tijdschrift voor Entomologie, 166*(2-3), 117–123. https://doi.org/10.1163/22119434-bja10025

Watanabe, R., & Yamasaki, S. (2020). Discovery of larvae of *Haliplus kamiyai* Nakane, 1963 (Coleoptera, Haliplidae) and Implications for its Life Cycle. *Elytra, Tokyo. New Series 10,* (2), 365–371.

Wichard, W., Arens, W., & Eisenbeis, G. (1995). *Atlas zur Biologie der Wasserinsekten* (XII + 338 pp). Stuttgart: Gustav Fischer.

Wilson, C. B. (1924). Water beetles in relation to pondfish culture, with life cycle histories of those found in fishponds at Fairport, Iowa. *Bulletin of the Bureau of Fisheries 39*, 1923–24. *Document, 953*, 231–345.

Yamasaki, S., Hirasawa, K., & Sano, S. (2020). Knowledge of larval and pupal period of *Haliplus eximius*. Sayabane. *New Series 40,* 17–18 [in Japanese].

Yamasaki, S. (2024). Knowledge of immature stages of *Haliplus angustifrons*. Sabayane. *New Series 55*, 16–17 [in Japanese].

Zaitzev, F. A. (1953). Coleoptera. Families Amphizoidae, Hygrobiidae, Haliplidae, Dytiscidae, Gyrinidae. *Fauna SSSR 4(N.S. 58),* 1–376. [In Russian; English translation 1972, Israel Program for Scientific Translations, Jerusalem.]

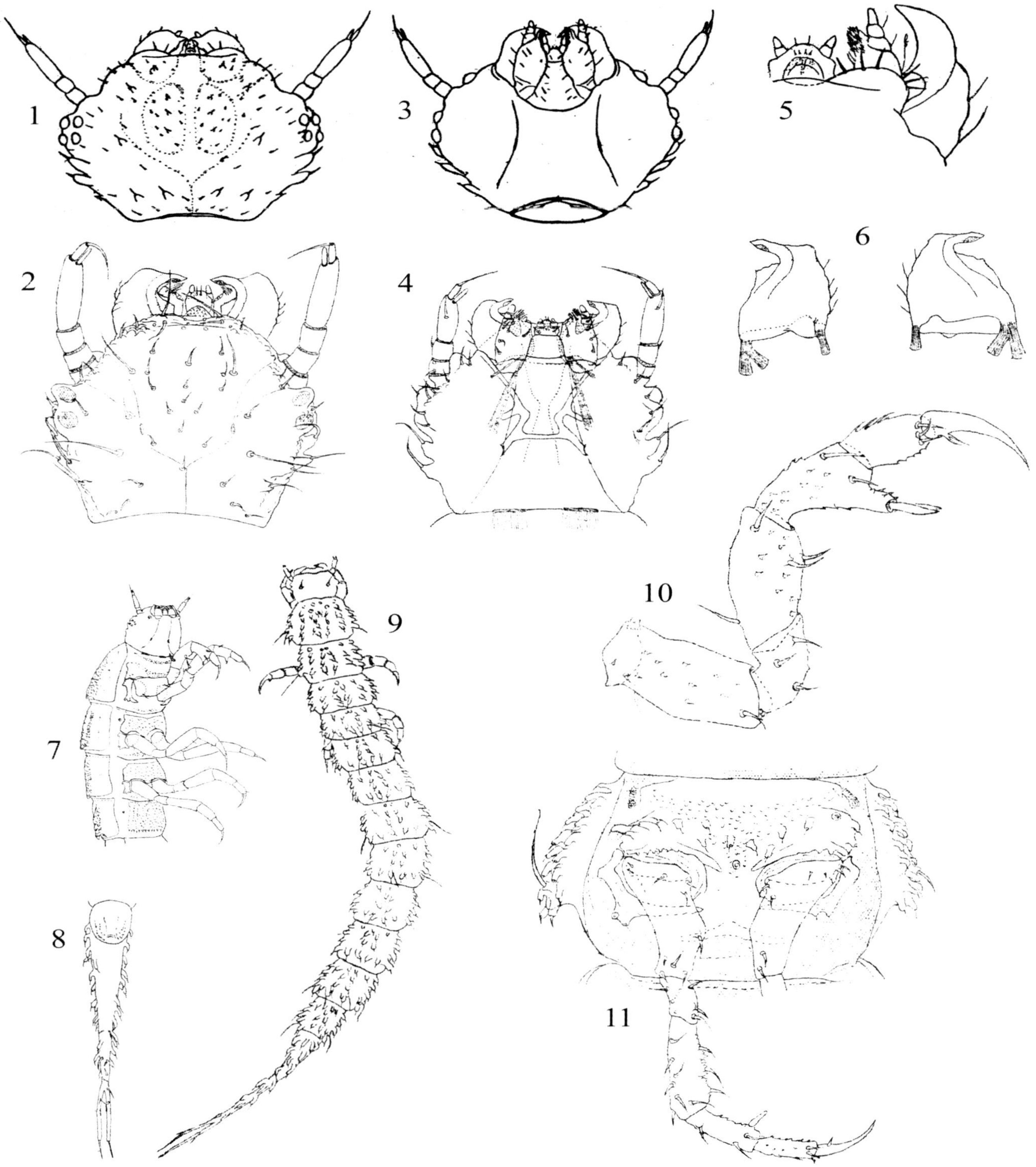

Plate 1. Figs. 1–11. *Haliplus*. General aspects of 3rd instar larva. 1, 2, head in dorsal view; 3, 4, Head in ventral view; 5, anterior region of the head in dorsal view; 6, left and right mandible; 7, head, thorax and abdominal segment I in lateral view; 8, abdominal segment X in ventral view; 9, larva in dorsal view; 10, foreleg; 11, mesothorax in ventral view. Figs. 1, 3, 5 after Bertrand (1928b). Figs. 2, 4, 6–11 after Jaboulet (1960).

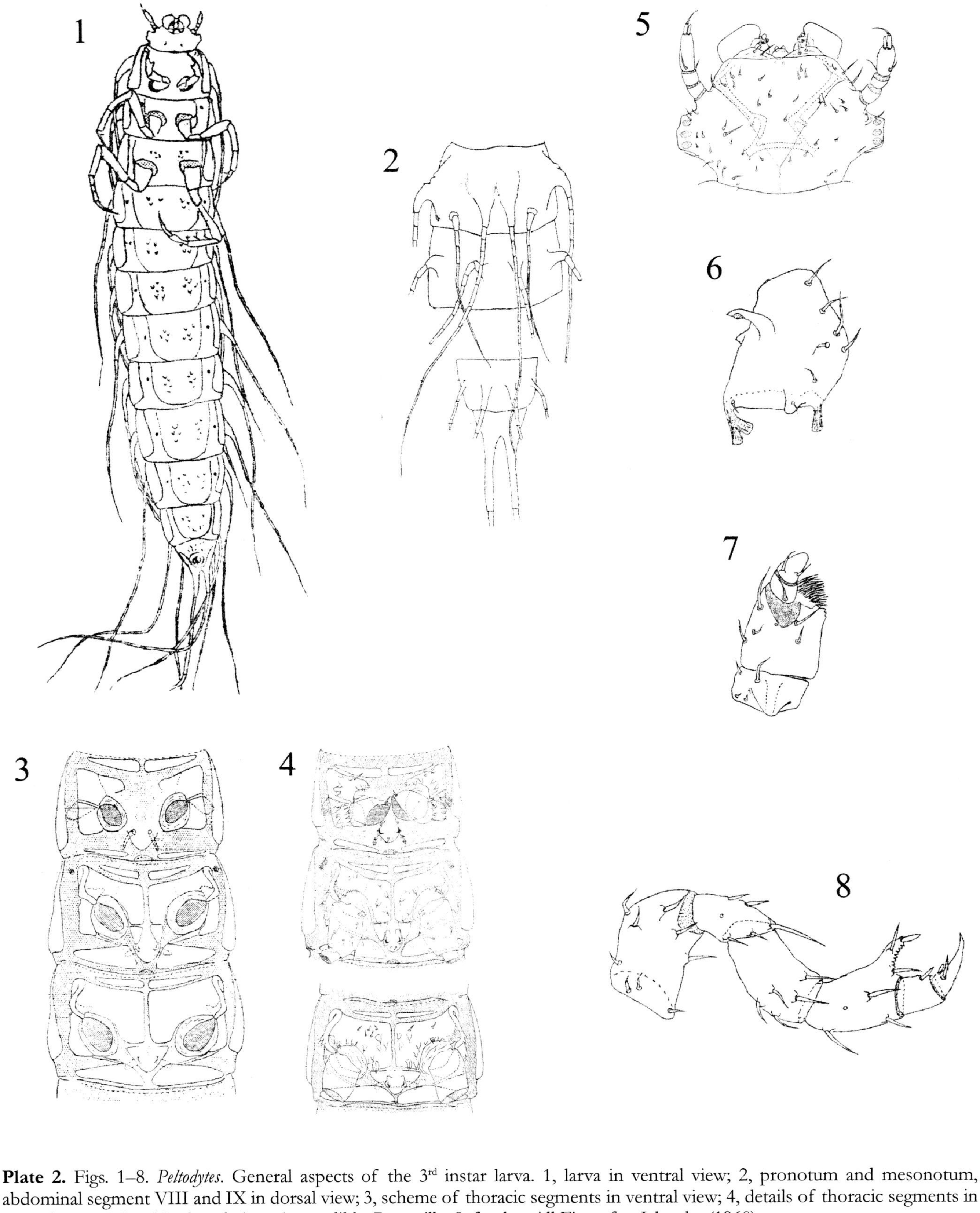

Plate 2. Figs. 1–8. *Peltodytes*. General aspects of the 3rd instar larva. 1, larva in ventral view; 2, pronotum and mesonotum, abdominal segment VIII and IX in dorsal view; 3, scheme of thoracic segments in ventral view; 4, details of thoracic segments in ventral view; 5, head in dorsal view; 6, mandible; 7, maxilla; 8, foreleg. All Figs. after Jaboulet (1960).

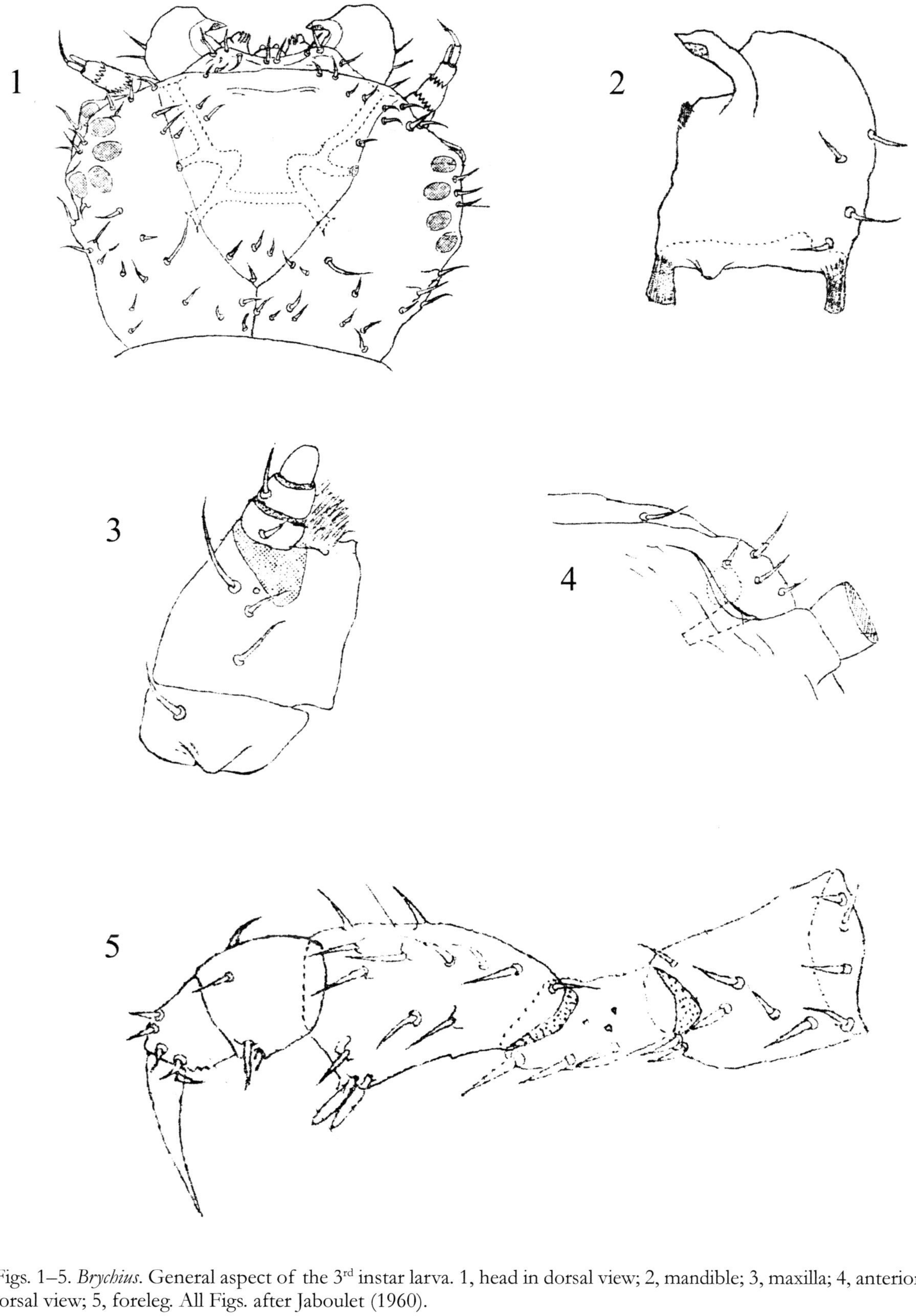

Plate 3. Figs. 1–5. *Brychius*. General aspect of the 3rd instar larva. 1, head in dorsal view; 2, mandible; 3, maxilla; 4, anterior part of head in dorsal view; 5, foreleg. All Figs. after Jaboulet (1960).

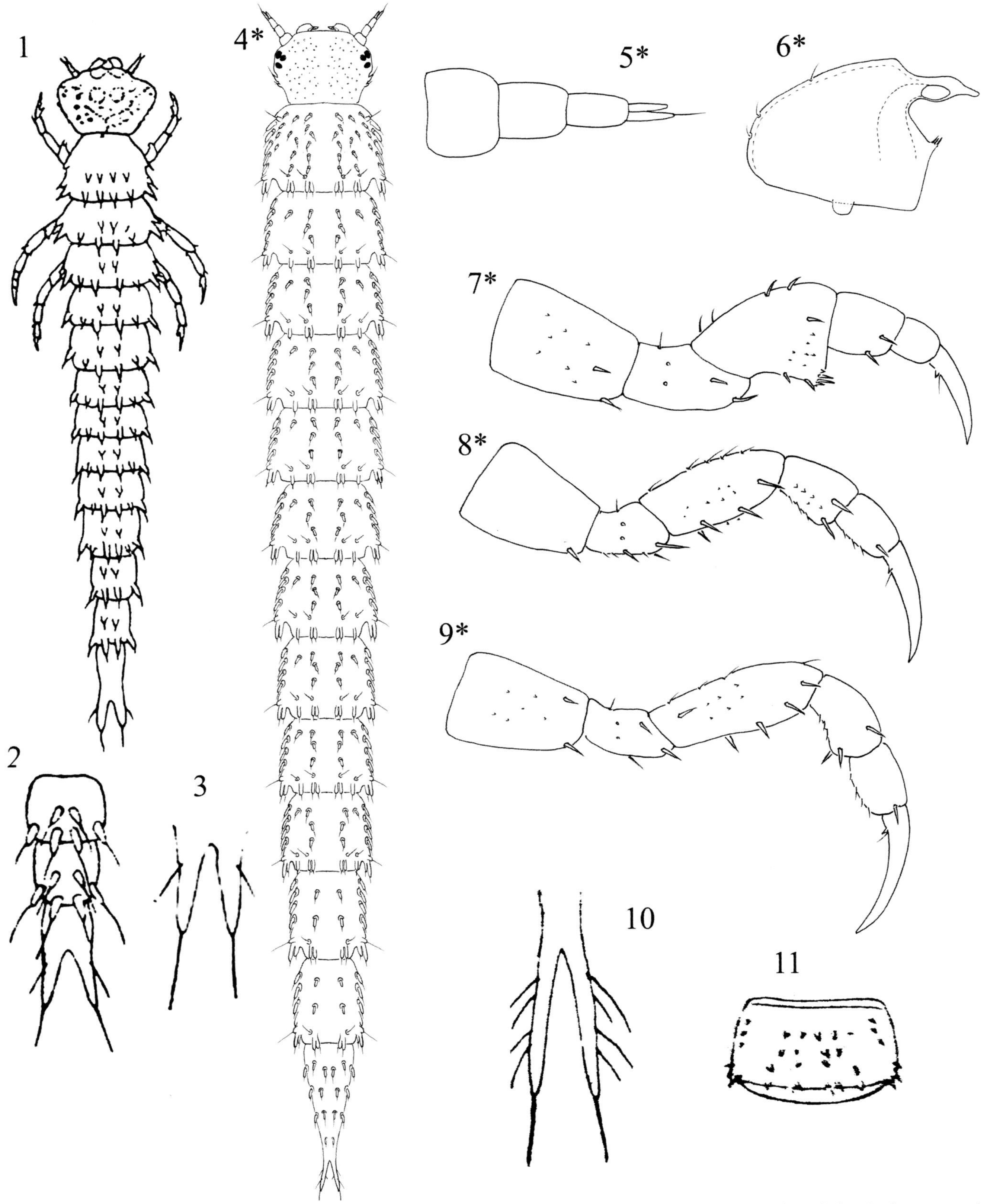

Plate 4. Figs. 1–11(*). *Brychius elevatus*. 1st and 2nd instar larva. 1–3, 1st instar: 1, larva in dorsal view; 2, last abdominal segments in ventral view; 3, forked part of segment X; 4–11, 2nd instar larva: 4*, body in dorsal view; 5*, antenna; 6*, mandible; 7*, foreleg, 8*; midleg; 9*, hindleg; 10, forked part of abdominal segment X; 11, abdominal segment I in dorsal view. Figs. 1–3, 10, 11 after Bertrand (1928b).

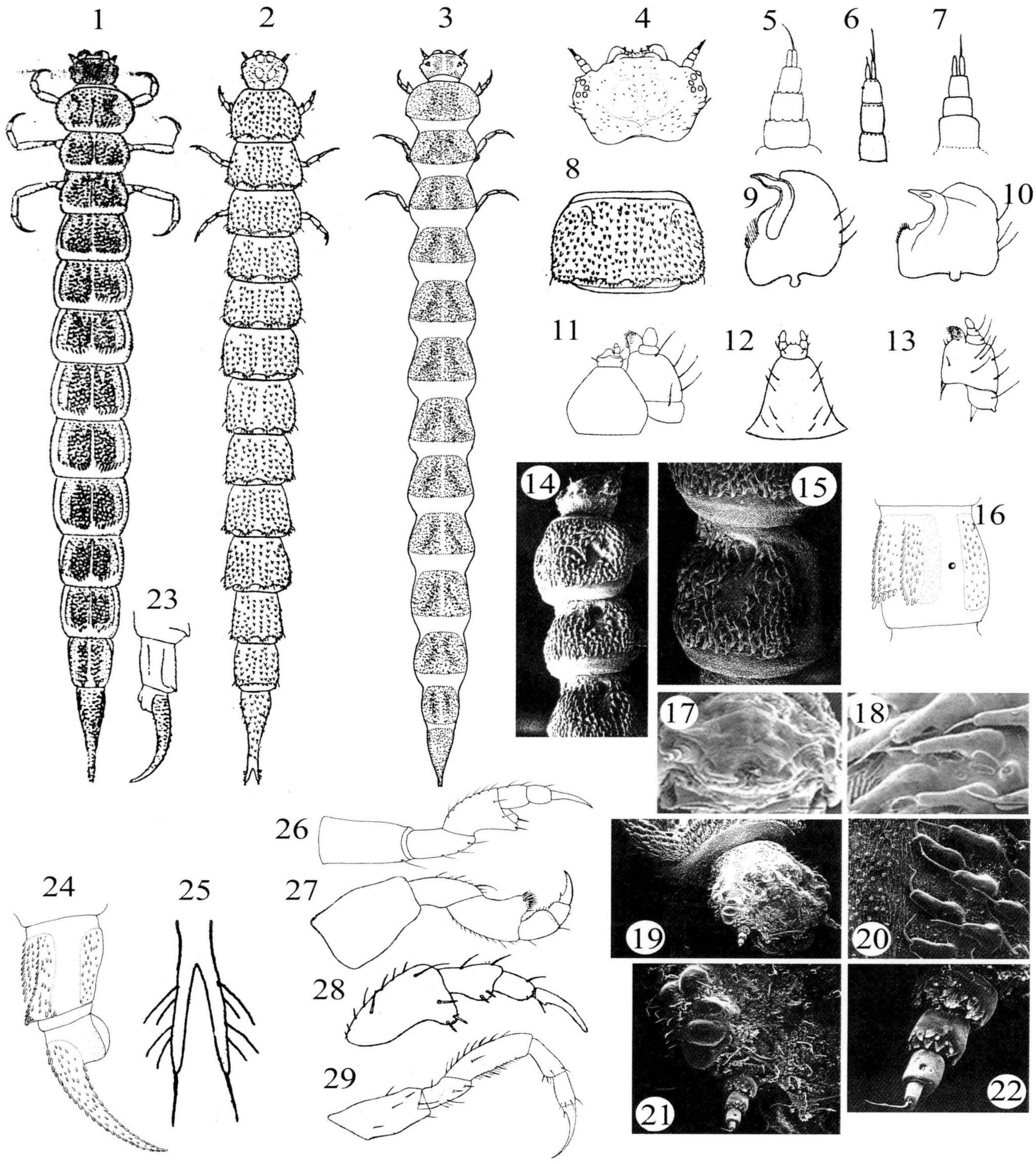

Plate 5. Figs. 1–29. *Brychius elevatus*. 3rd instar larva. 1–3, larva in dorsal view; 4, head in dorsal view; 5–7, antenna; 8, abdominal segment I in dorsal view; 9,10, mandible; 11, labium and maxilla; 12, labium; 13, maxilla; 14, head and thorax in dorsal view; 15, abdominal segment I in dorsal view; 16, abdominal segment in lateral view; 17, head in frontal view; 18, 20, tubercles on dorsal side of thoracic and abdominal segments; 19, 21, head in oblique frontal view; 22, antenna in dorsal view; 23, 24, abdominal segments IX and X in lateral view; 25, forked part of abdominal segment X; 26–28, foreleg; 29, hindleg; Figs. 1, 5, 9, 11, 23, 27, 29, after Rousseau (1919a). Figs. 2, 4, 6, 8, 10, 12, 13, 25, 28, after Bertrand (1928b). Figs. 14, 15, 19–22 aftter Wichard et al. (1995). Figs. 3, 7, 16, 24, 26 after Vondel (1997). Figs. 17, 18 after Vondel (2016).

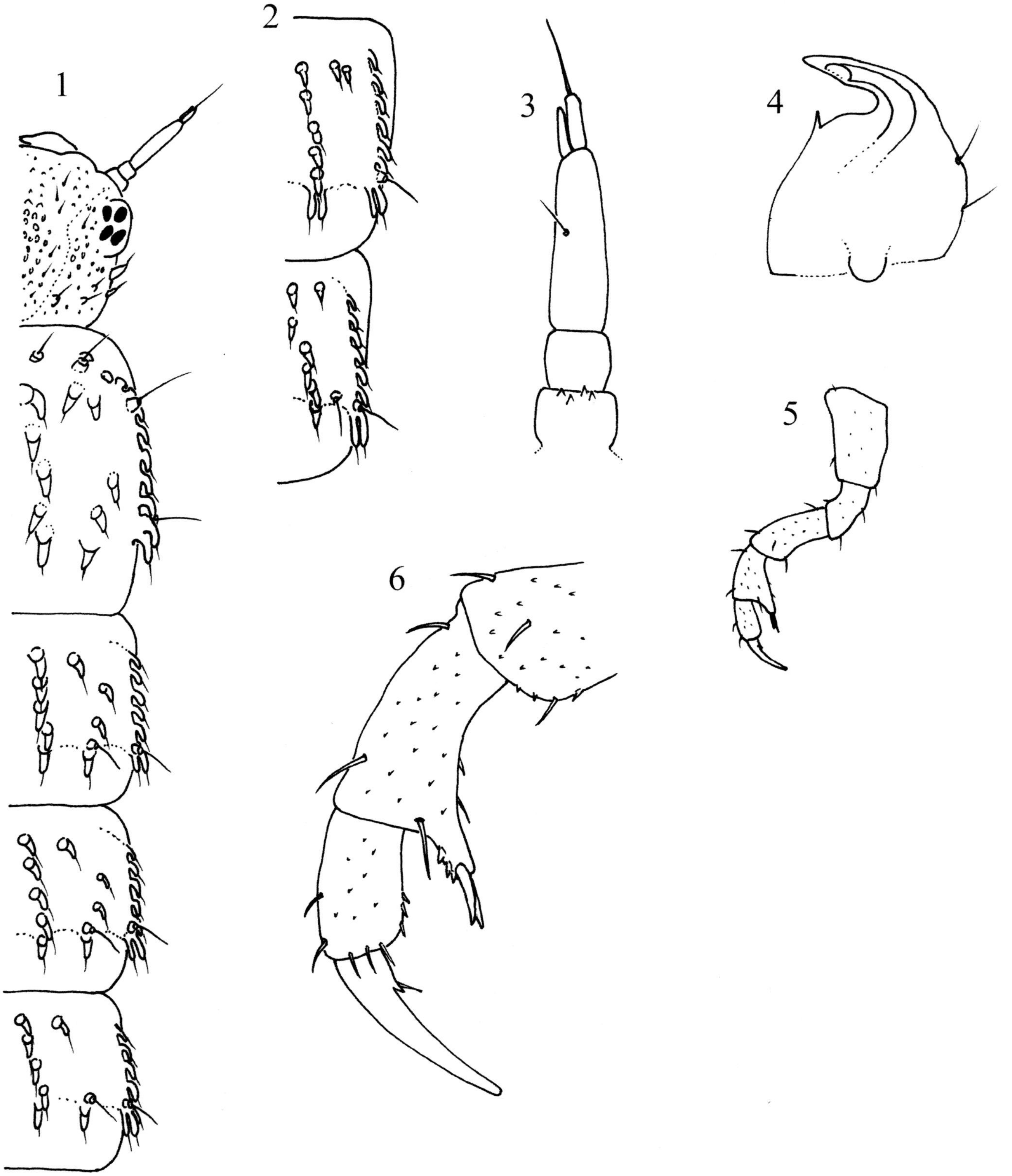

Plate 6. Figs. 1–6. *Haliplus (Haliaplus) laminatus.* 2nd instar larva. 1, head, thorax and first abdominal segment in dorsal view (right half); 2, abdominal segment VIII and IX in dorsal view (right half); 3, antenna; 4, mandible; 5, foreleg; 6, claw, tarsus and tibia of foreleg. All Figs. after Vondel (1986).

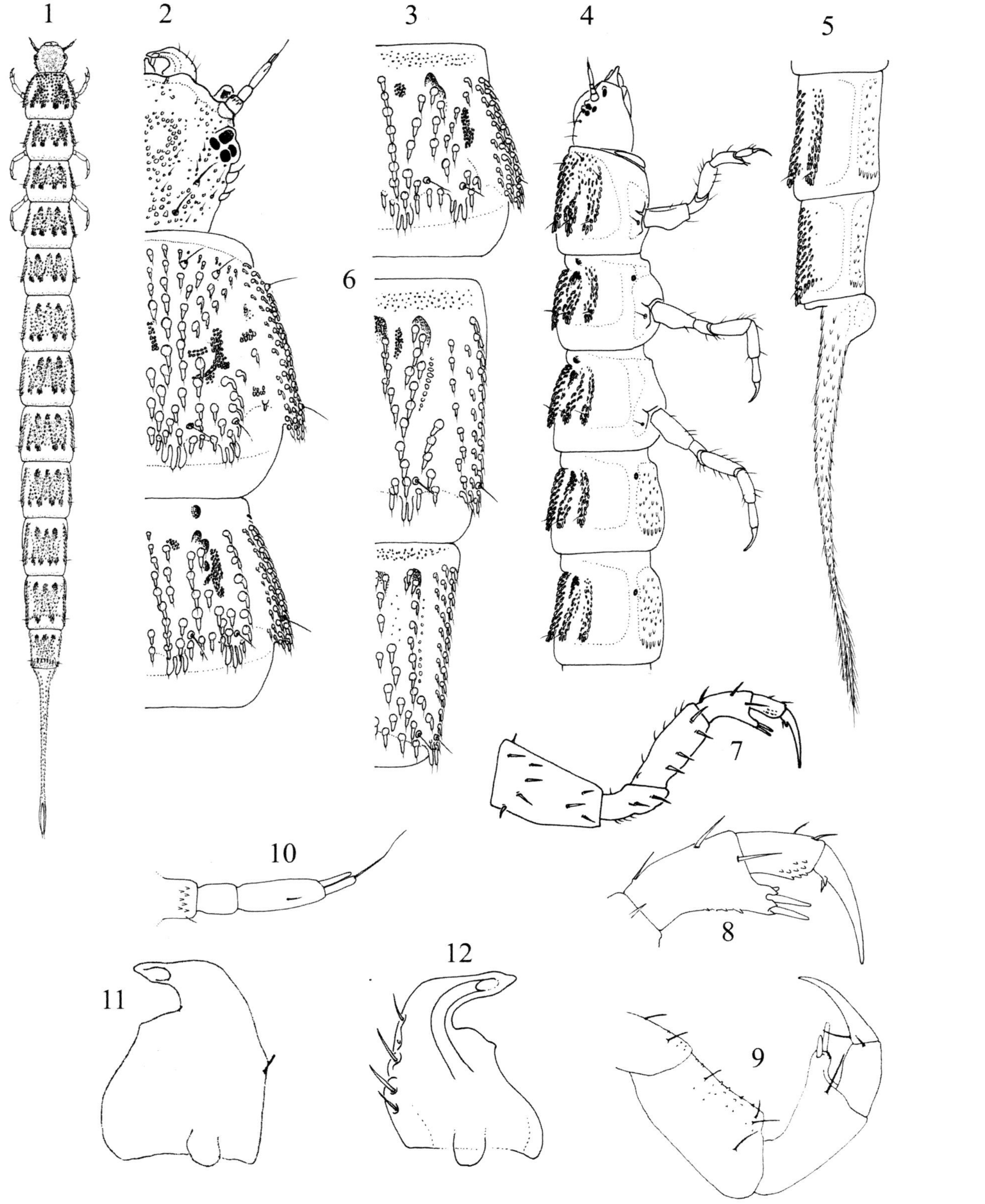

Plate 7. Figs. 1–12. *Haliplus (Haliaplus) laminatus*. 3rd instar larva. 1, body in dorsal view; 2, head, pronotum and mesonotum in dorsal view (right half); 3, abdominal segment I in dorsal view (right half); 4, head, thoracic segments and first two abdominal segments in lateral view, 5, last three abdominal segments in lateral view; 6, abdominal segment VIII an IX in dorsal view (right half); 7, foreleg; 8, claw, tarsus and tibia of foreleg; 9, claw, tarsus, tibia and femur of foreleg; 10, antenna; 11, 12, mandible. Figs. 9, 11 after Holmen (1981). Other Figs. after Vondel (1986).

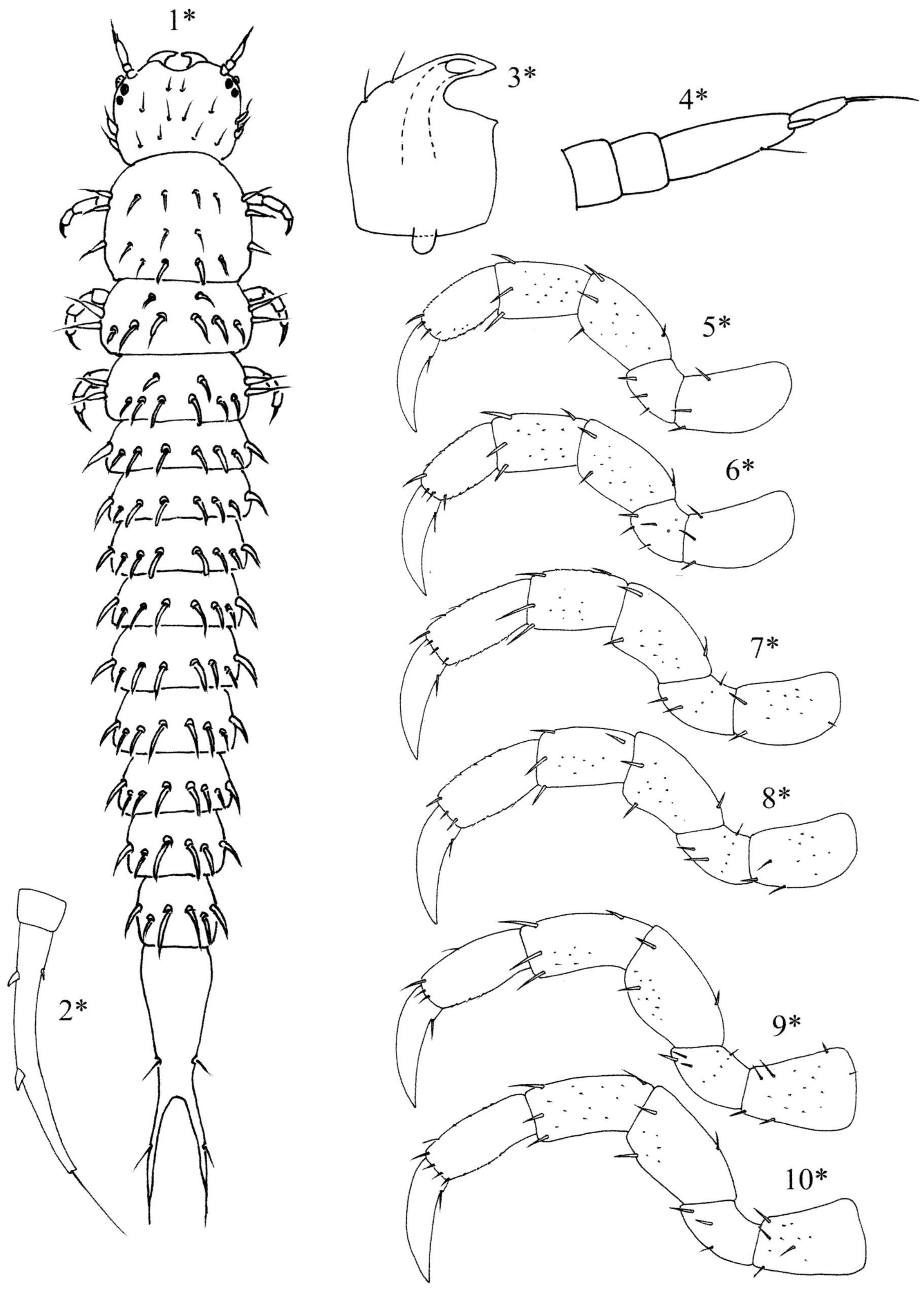

Plate 8. Figs. 1*–10*. *Haliplus (Haliplidius) confinis*. 1st instar larva. 1*, body in dorsal view; 2*, dorsal tubercle; 3*, mandible; 4*, antenna; 5*, foreleg in dorsal view; 6*, foreleg in ventral view; 7*, midleg in dorsal view; 8*, midleg in ventral view; 9*, hindleg in dorsal view: 10*, hindleg in ventral view.

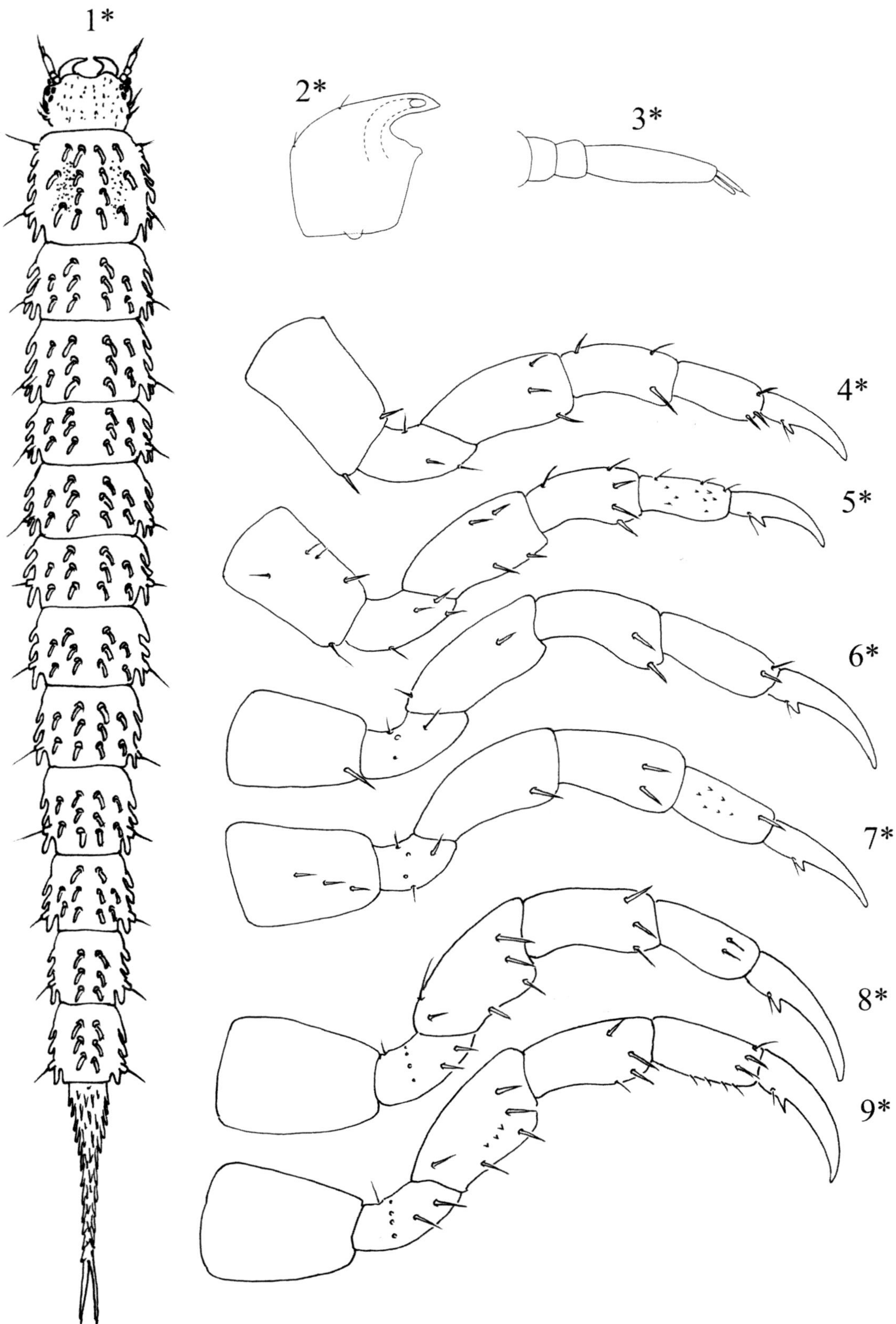

Plate 9. Figs. 1*–9*. *Haliplus (Haliplidius) confinis*. 2nd instar larva. 1*, body in dorsal view; 2*, mandible; 3*, antenna; 4*, foreleg in dorsal view: 5*, foreleg in ventral view; 6*, midleg in dorsal view; 7*, midleg in ventral view; 8*, hindleg in dorsal view; 9*, hindleg in ventral view.

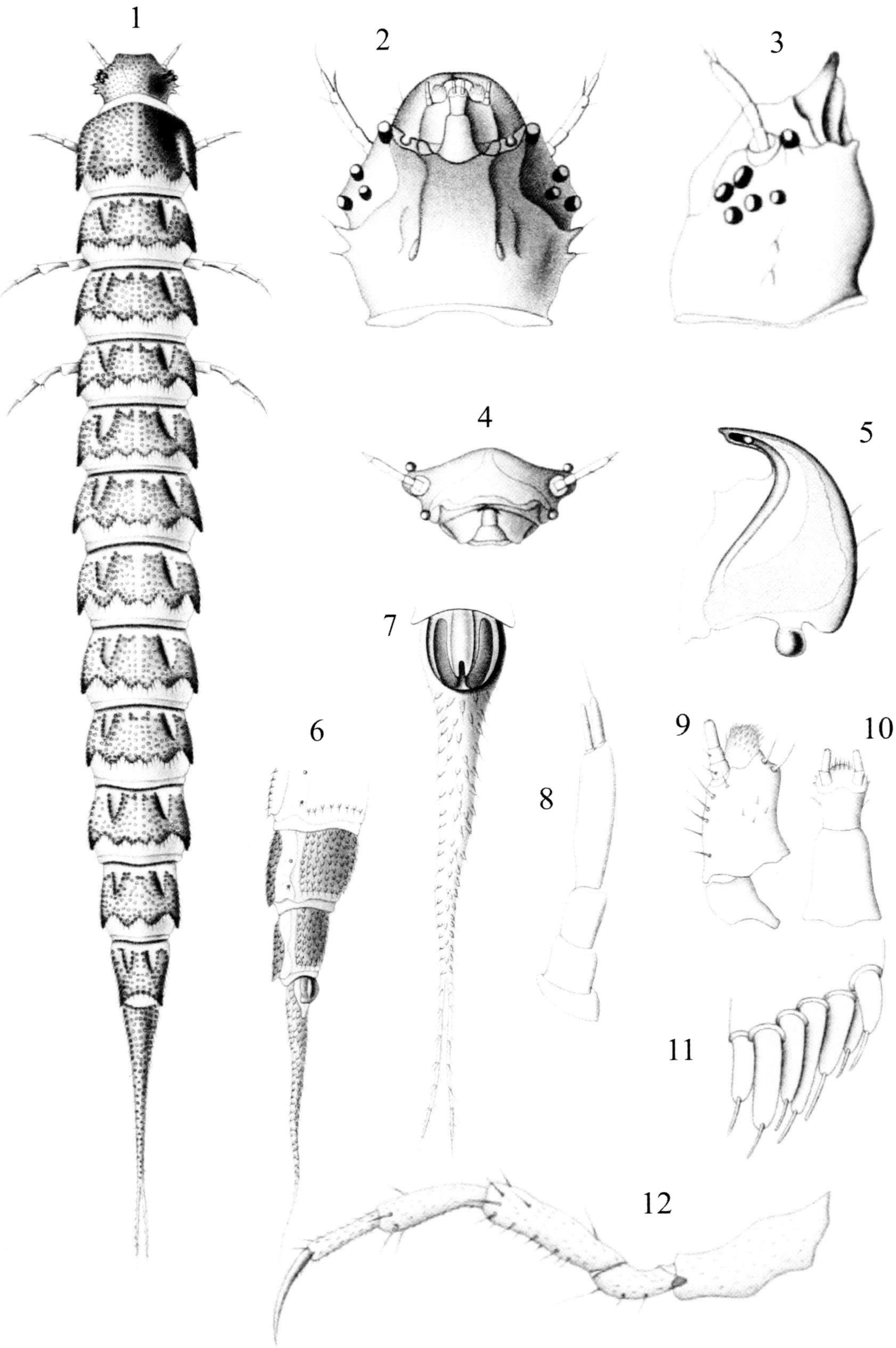

Plate 10. Figs. 1–12. *Haliplus (Haliplidius) confinis.* 3rd instar larva. 1, body in dorsal view; 2, head in dorsal view; 3, head in lateral view; 4, head in frontal view; 5, mandible; 6, last three abdominal segments in lateral view; 7, abdominal segment X in ventral view; 8, antenna; 9, maxilla; 10, labium; 11, lateral tubercles; 12, hindleg.. All Figs. after Schiødte (1864).

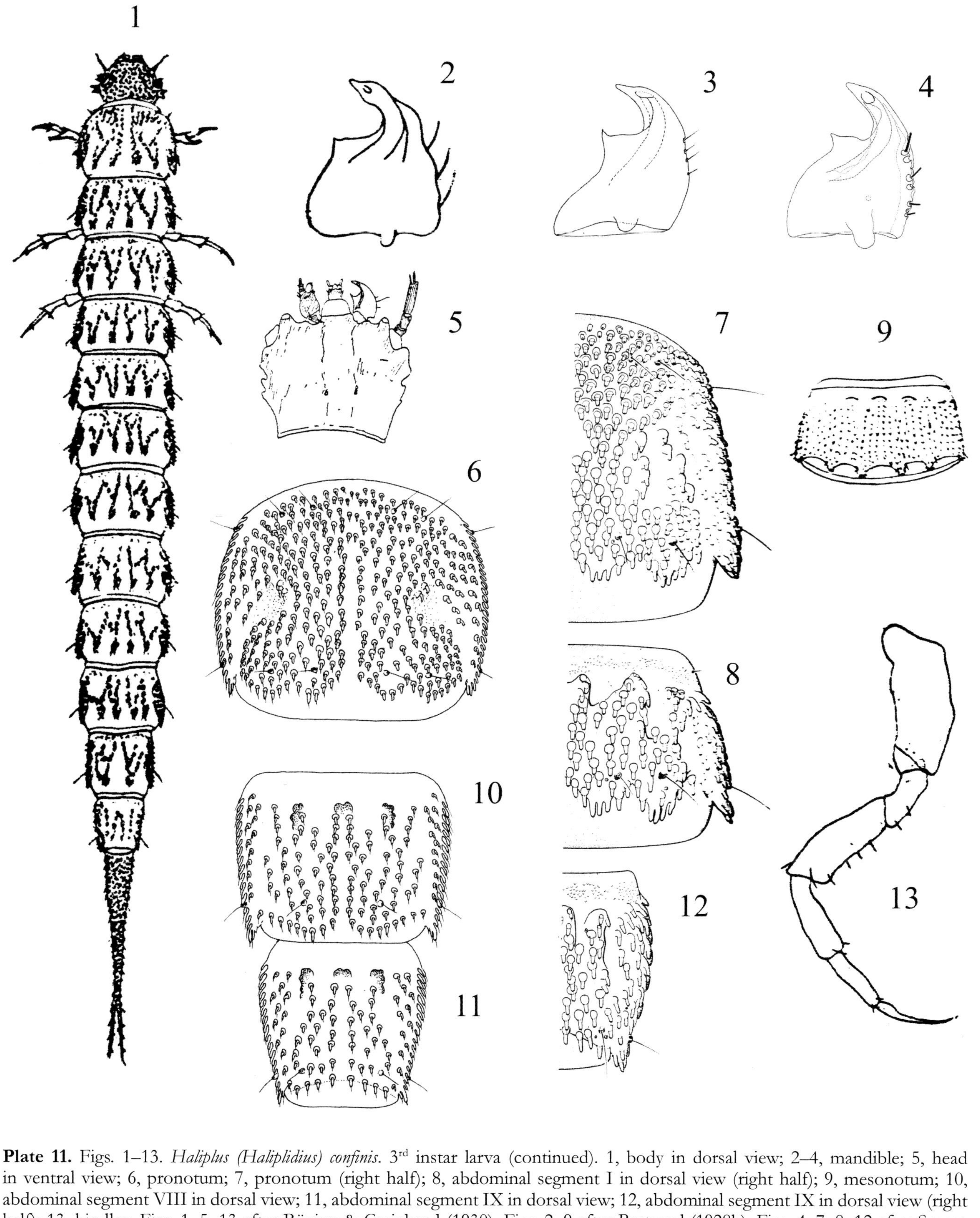

Plate 11. Figs. 1–13. *Haliplus (Haliplidius) confinis*. 3rd instar larva (continued). 1, body in dorsal view; 2–4, mandible; 5, head in ventral view; 6, pronotum; 7, pronotum (right half); 8, abdominal segment I in dorsal view (right half); 9, mesonotum; 10, abdominal segment VIII in dorsal view; 11, abdominal segment IX in dorsal view; 12, abdominal segment IX in dorsal view (right half); 13, hindleg. Figs. 1, 5, 13 after Böving & Craighead (1930). Figs. 2, 9 after Bertrand (1928b). Figs. 4, 7, 8, 12 after Seeger (1971a). Figs. 3, 6, 10, 11 after Vondel (1997).

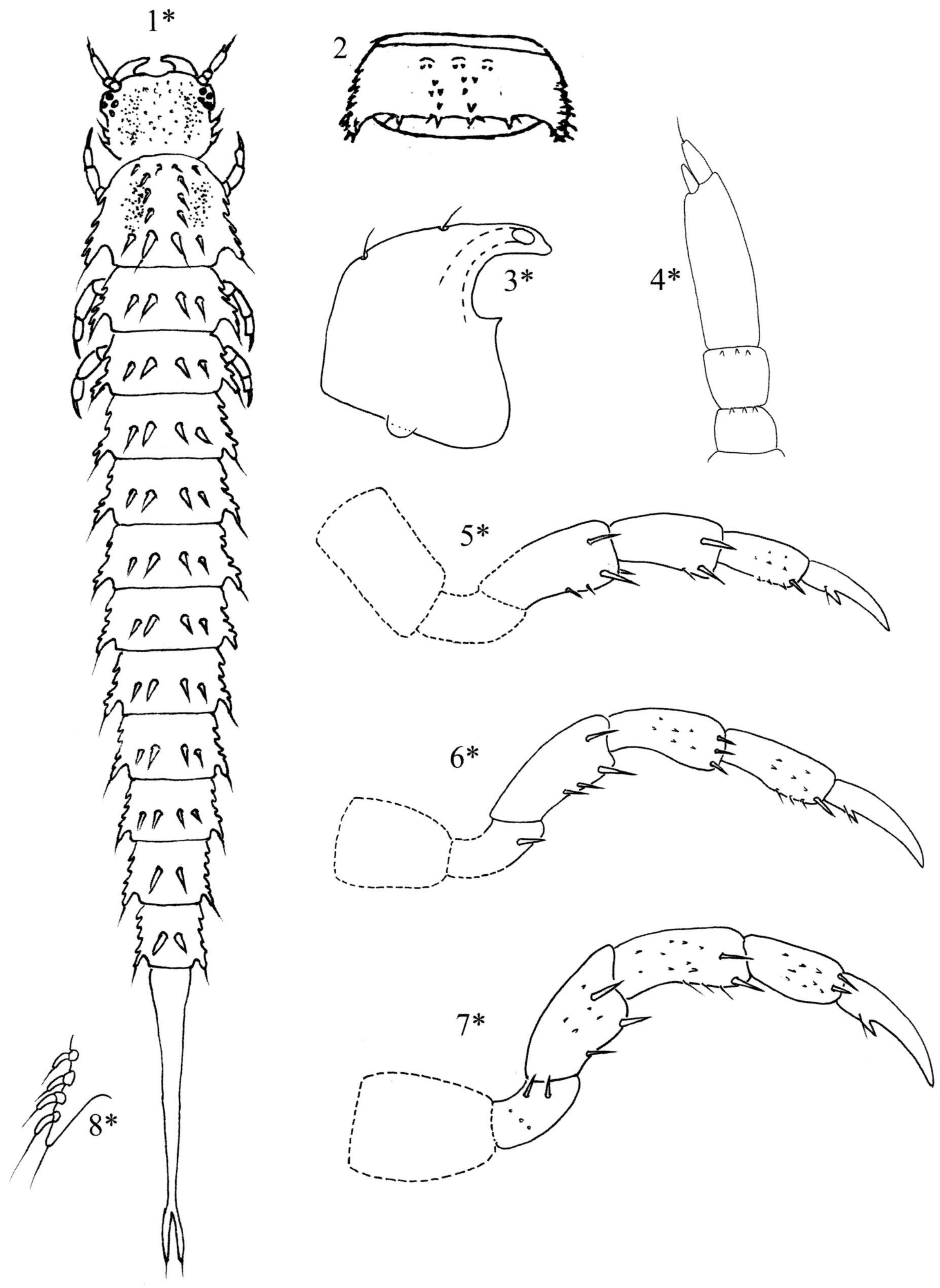

Plate 12. Figs. 1–8(*). *Haliplus (Haliplidius) obliquus*. 2nd instar larva. 1*, body in dorsal view; 2, abdominal segment I in dorsal view; 3*, mandible; 4*, antenna; 5*, foreleg; 6*, midleg; 7*, hindleg; 8*, lateral extension of thoracic and abdominal segments. Fig. 2 after Bertrand (1928b).

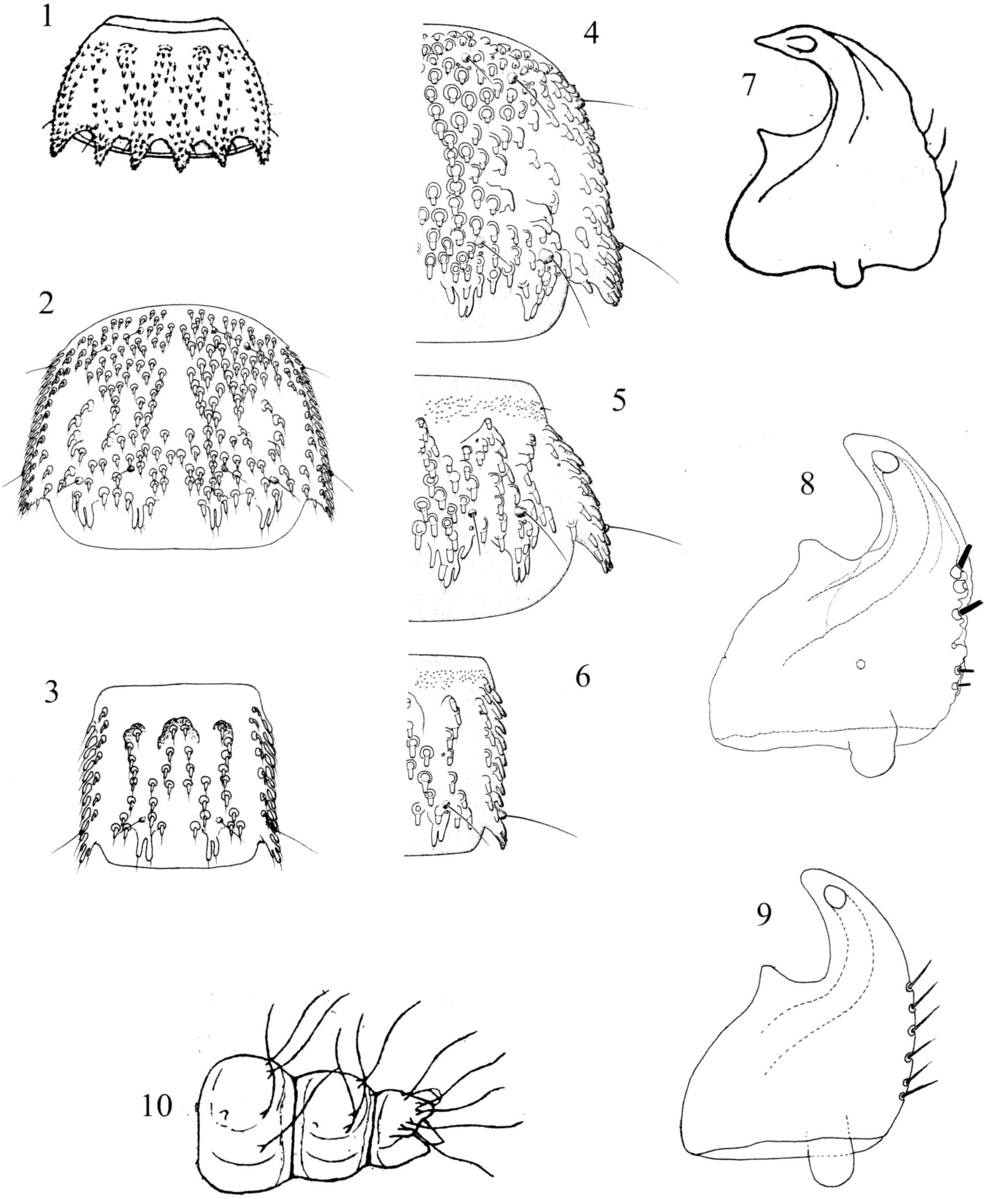

Plate 13. Figs. 1–10. *Haliplus (Haliplidius) obliquus*. 3rd instar larva and pupa. 1, abdominal segment I in dorsal view; 2, pronotum; 3, abdominal segment IX in dorsal view; 4, pronotum (right half); 5, abdominal segment I in dorsal view (right half); 6, abdominal segment IX in dorsal view (right half); 7–9, mandible; 10, pupa. Figs. 1, 7, 10 after Bertrand (1928b). Figs. 4–6, 8 after Seeger (1971a). Figs. 2, 3, 9 after Vondel (1997).

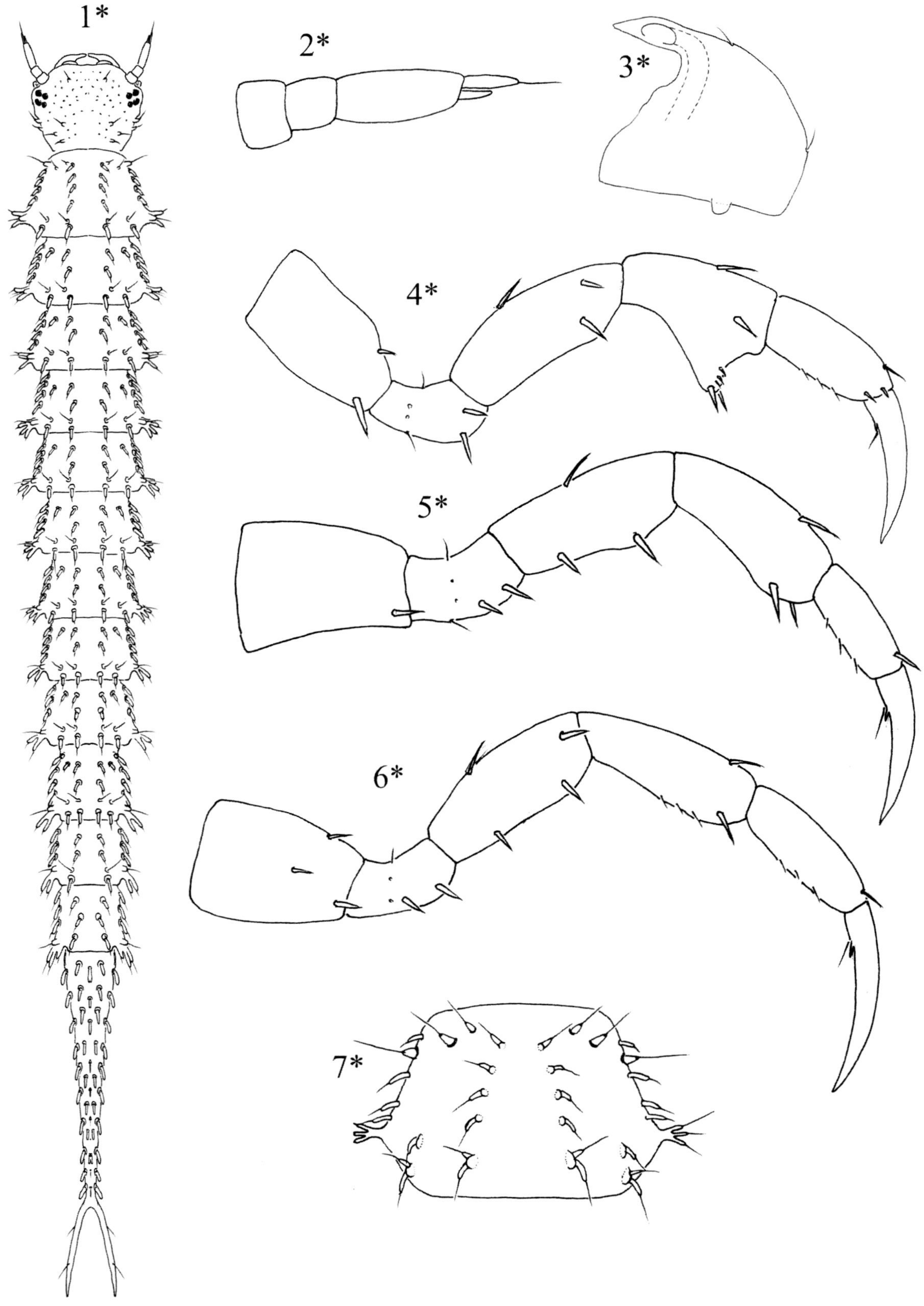

Plate 14. Figs. 1*–7*. *Haliplus (Haliplidius) varius*. 2nd instar larva. 1*, body in dorsal view; 2*, antenna; 3*, mandible; 4*, foreleg; 5*, midleg; 6*, hindleg; 7*, pronotum.

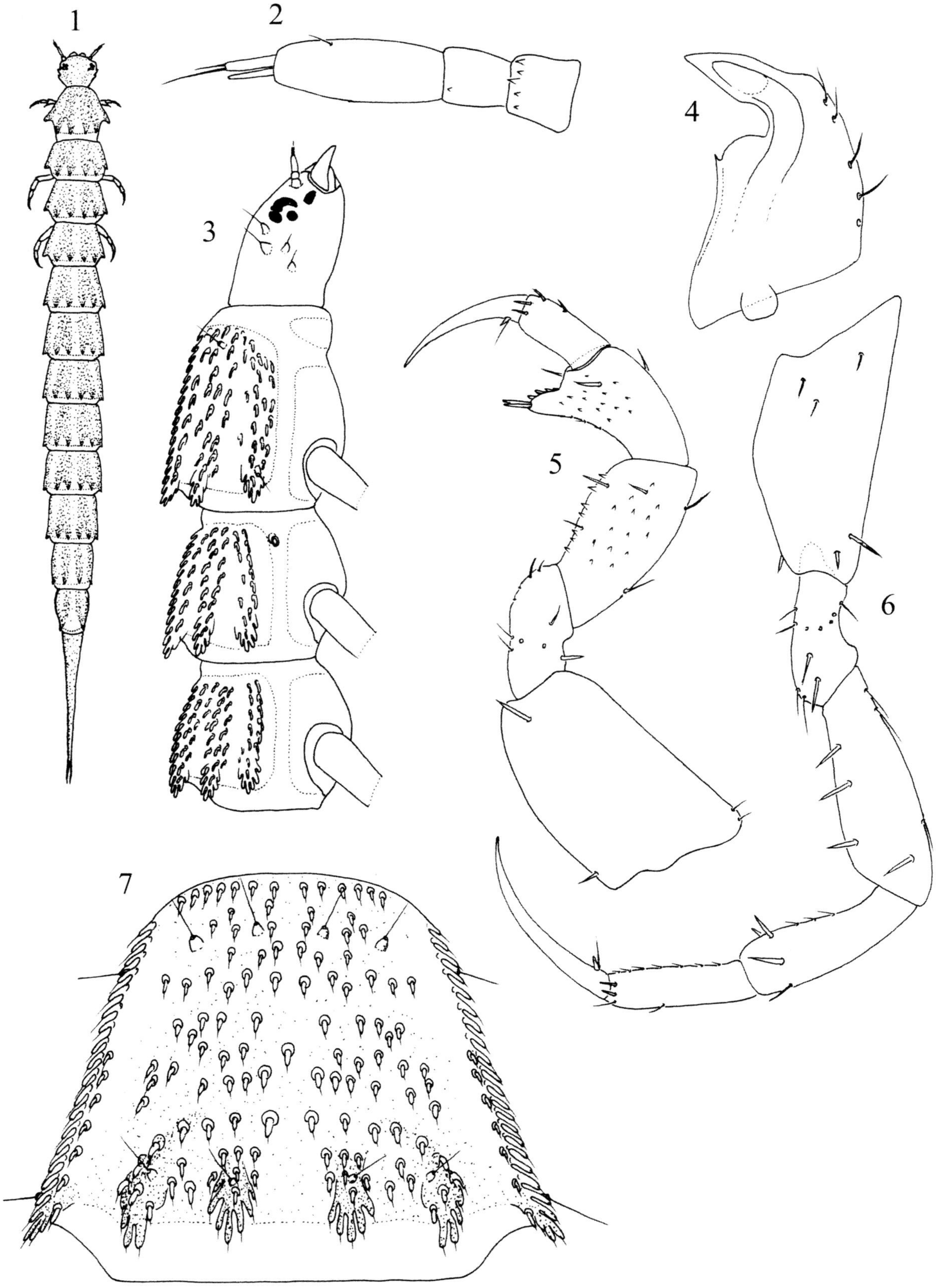

Plate 15. Figs. 1–7. *Haliplus (Haliplidius) varius.* 3rd instar larva. 1, larva in dorsal view; 2, antenna; 3, head and thoracic segments in lateral view; 4, mandible; 5, foreleg; 6, hindleg; 7, pronotum. All Figs. after Vondel (1996).

1

2

Plate 16. Figs. 1–2. *Haliplus (Haliplus?)* sp. from Japan. 1, 2nd instar larva; 2, 3rd instar larva. Both Figs. after Kitano & Sano (2011).

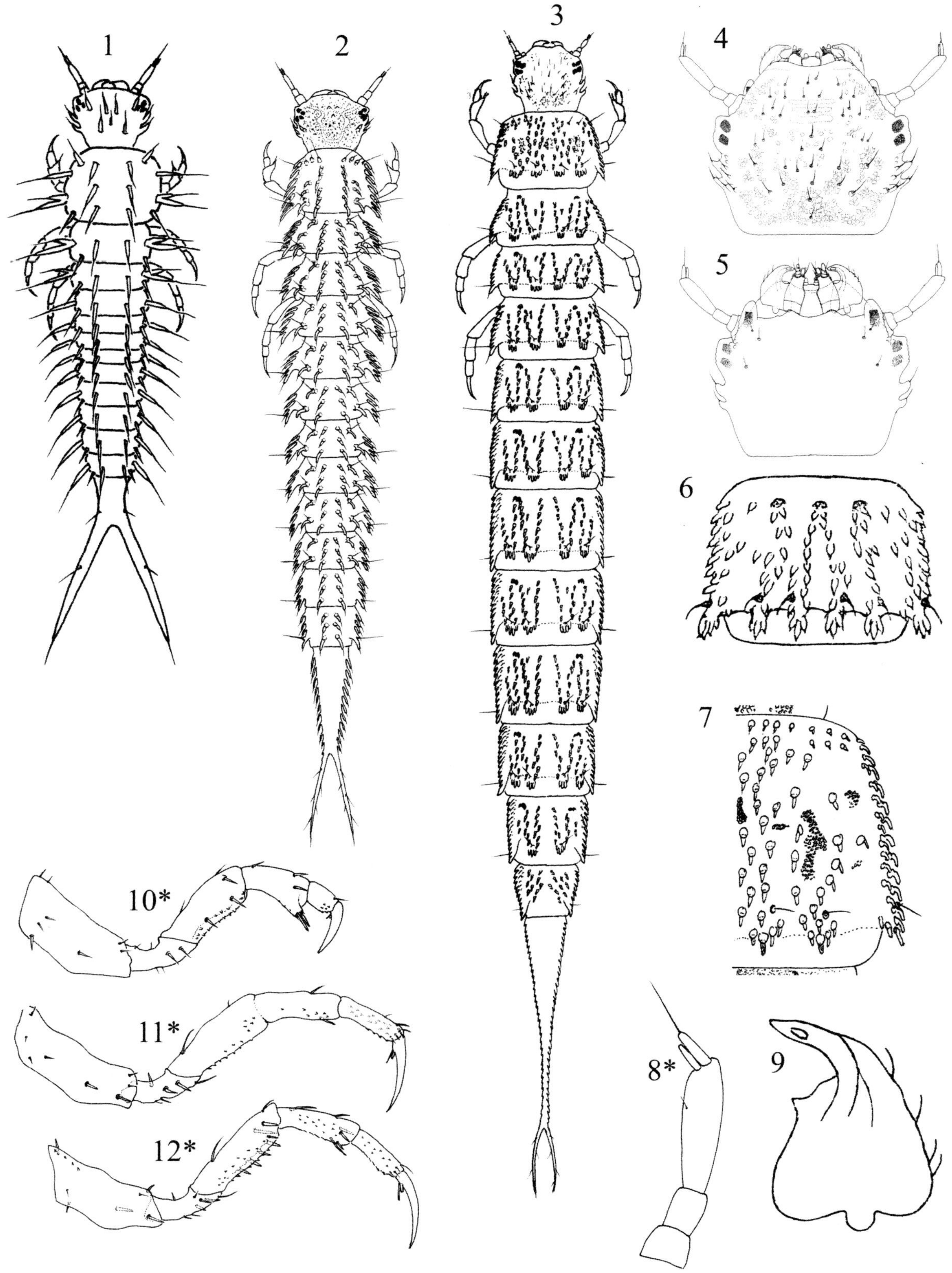

Plate 17. Figs. 1–12(*). *Haliplus (Haliplus) apicalis.* 1st, 2nd and 3rd instar larva. 1, 1st instar larva in dorsal view. 2, 2nd instar larva in dorsal view. Figs. 3–12, 3rd instar larva: 3, larva in dorsal view; 4, head in dorsal view; 5, head in ventral view; 6, abdominal segment I in dorsal view; 7, pronotum (right half); 8* antenna; 9, mandible; 10*, foreleg; 11*, midleg; 12*, hindleg. Figs. 1–5, 7 after Vondel (1997). Fig. 6, 9 after Bertrand (1942).

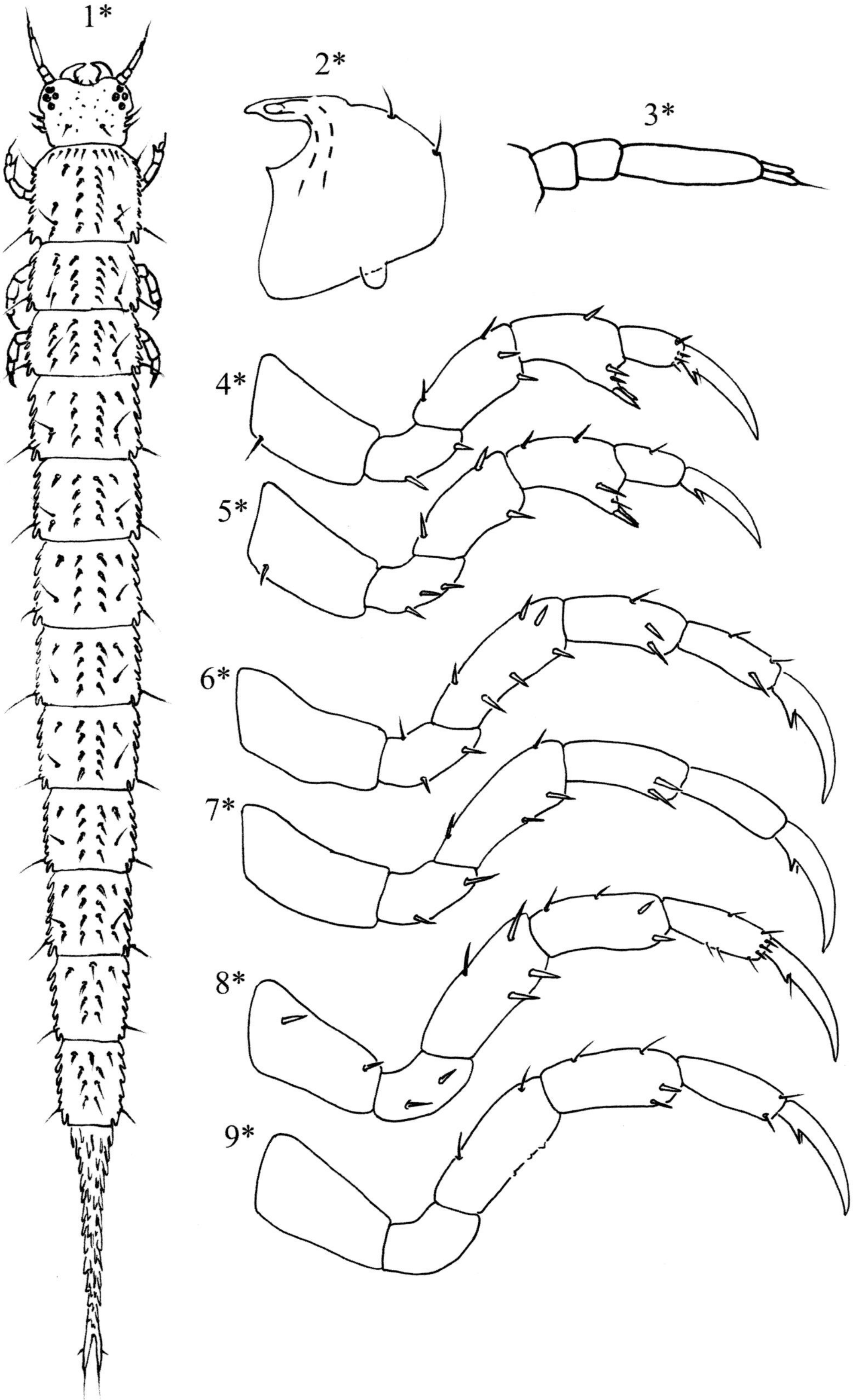

Plate 18. Figs. 1*–9*. *Haliplus (Haliplus) fluviatilis*. 2nd instar larva. 1*, larva in dorsal view; 2*, mandible; 3*, antenna; 4*, foreleg in dorsal view; 5*, foreleg in ventral view; 6*, midleg in dorsal view; 7*, midleg in ventral view; 8*, hindleg in dorsal view; 9*, hindleg in ventral view.

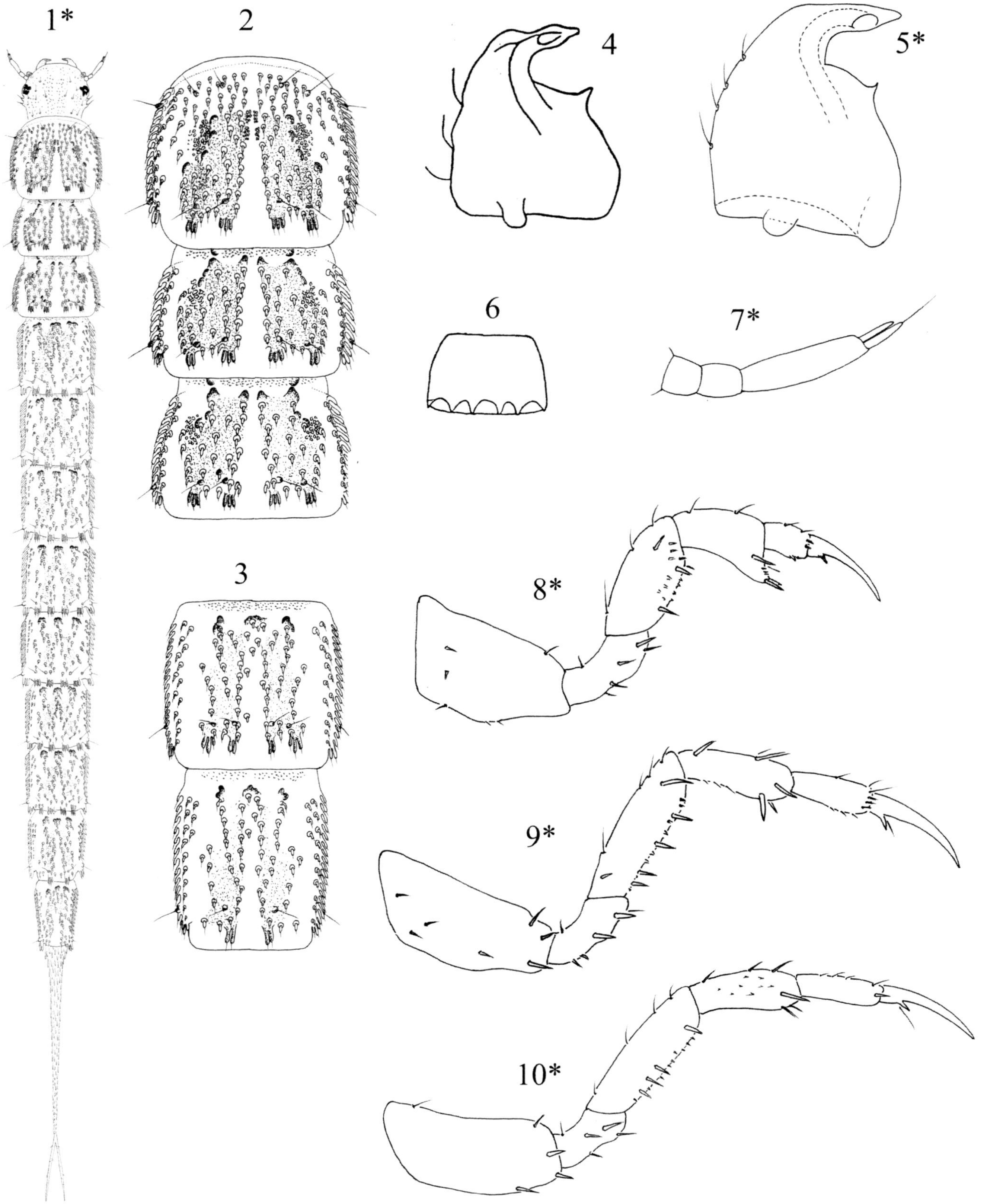

Plate 19. Figs. 1–10(*). *Haliplus (Haliplus) fluviatilis*. 3rd instar larva. 1*, body in dorsal view; 2, thorax in dorsal view; 3, abdominal segment VII and VIII in dorsal view; 4, 5*, mandible; 6, abdominal segment I in dorsal view; 7*, antenna; 8*, foreleg; 9*, midleg, 10*, hindleg. Figs. 4, 6 after Bertrand (1950). Figs. 2, 3 after Vondel (1997).

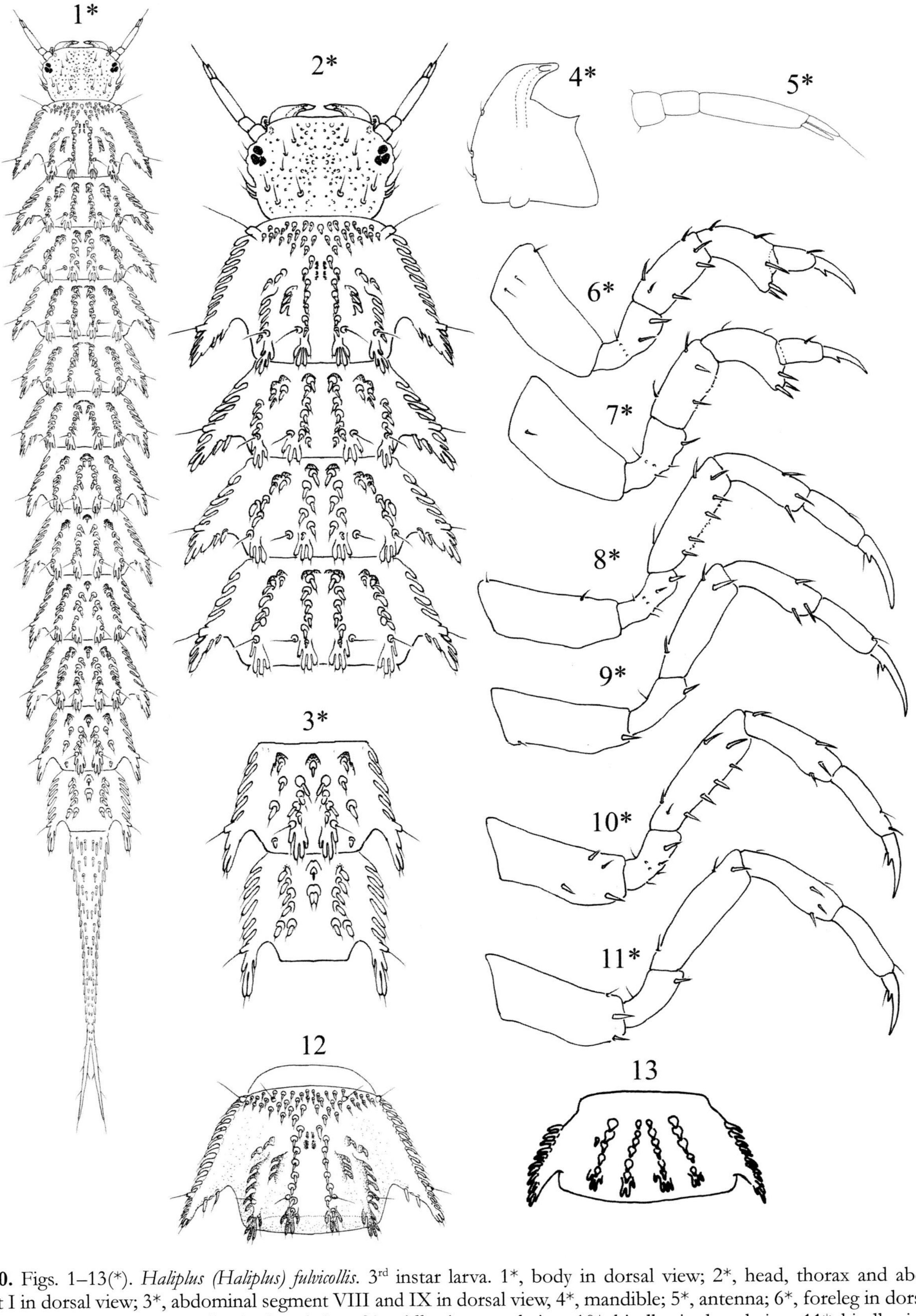

Plate 20. Figs. 1–13(*). *Haliplus (Haliplus) fulvicollis*. 3rd instar larva. 1*, body in dorsal view; 2*, head, thorax and abdominal segment I in dorsal view; 3*, abdominal segment VIII and IX in dorsal view, 4*, mandible; 5*, antenna; 6*, foreleg in dorsal view; 7*, foreleg in ventral view; 8*, midleg in dorsal view; 9*, midleg in ventral view; 10*, hindleg in dorsal view; 11*, hindleg in ventral view; 12, pronotum; 13, abdominal segment I in dorsal view. Fig. 13 after Holmen (1987). Fig. 12 after Vondel (1997).

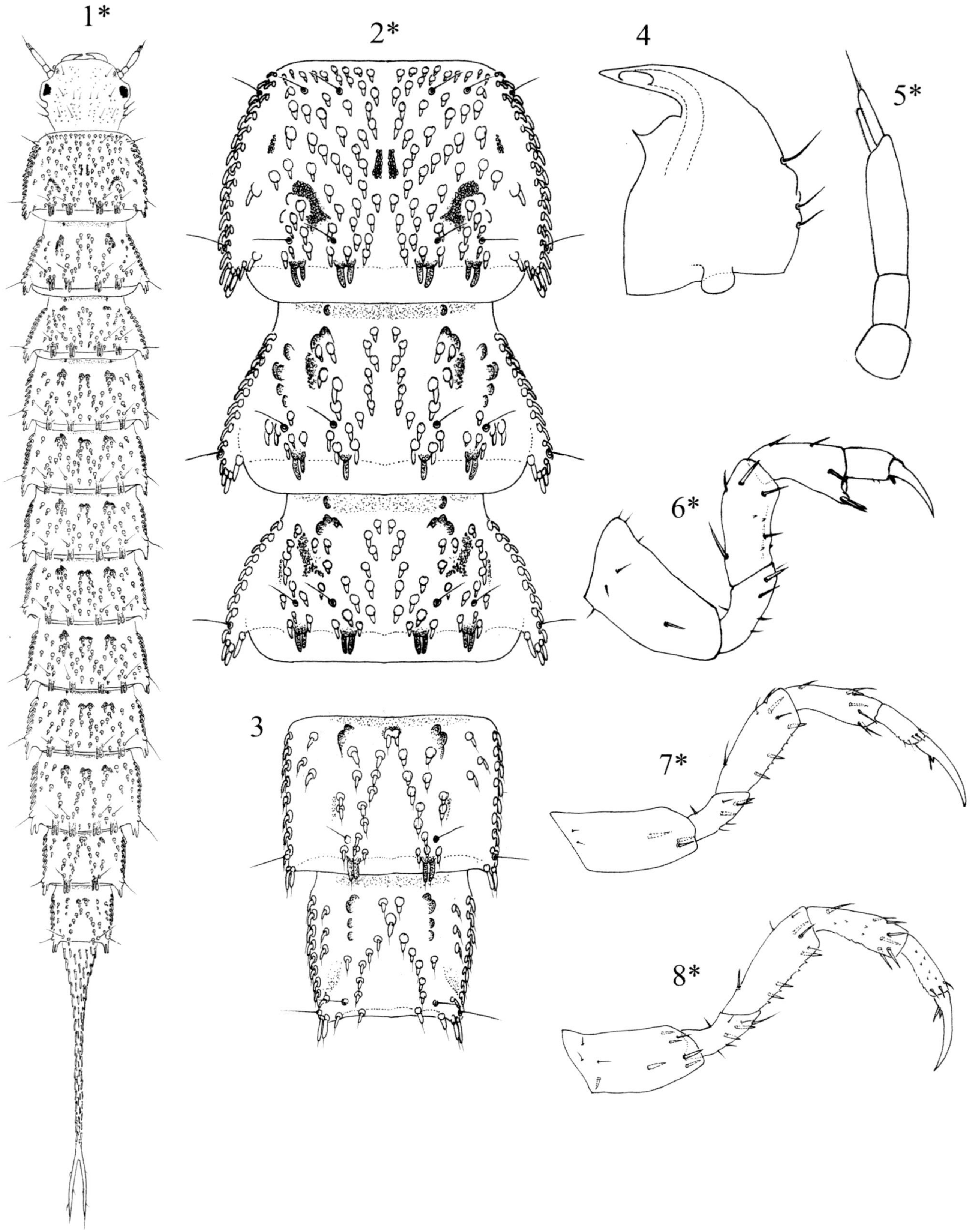

Plate 21. Figs. 1–8(*). *Haliplus (Haliplus) furcatus*. 3rd instar larva. 1*, body in dorsal view; 2*, thorax in dorsal view; 3, abdominal segment VIII and IX in dorsal view; 4, mandible; 5*, antenna; 6*, foreleg; 7*, midleg; 8*, hindleg. Figs. 3, 4 after Vondel (1997).

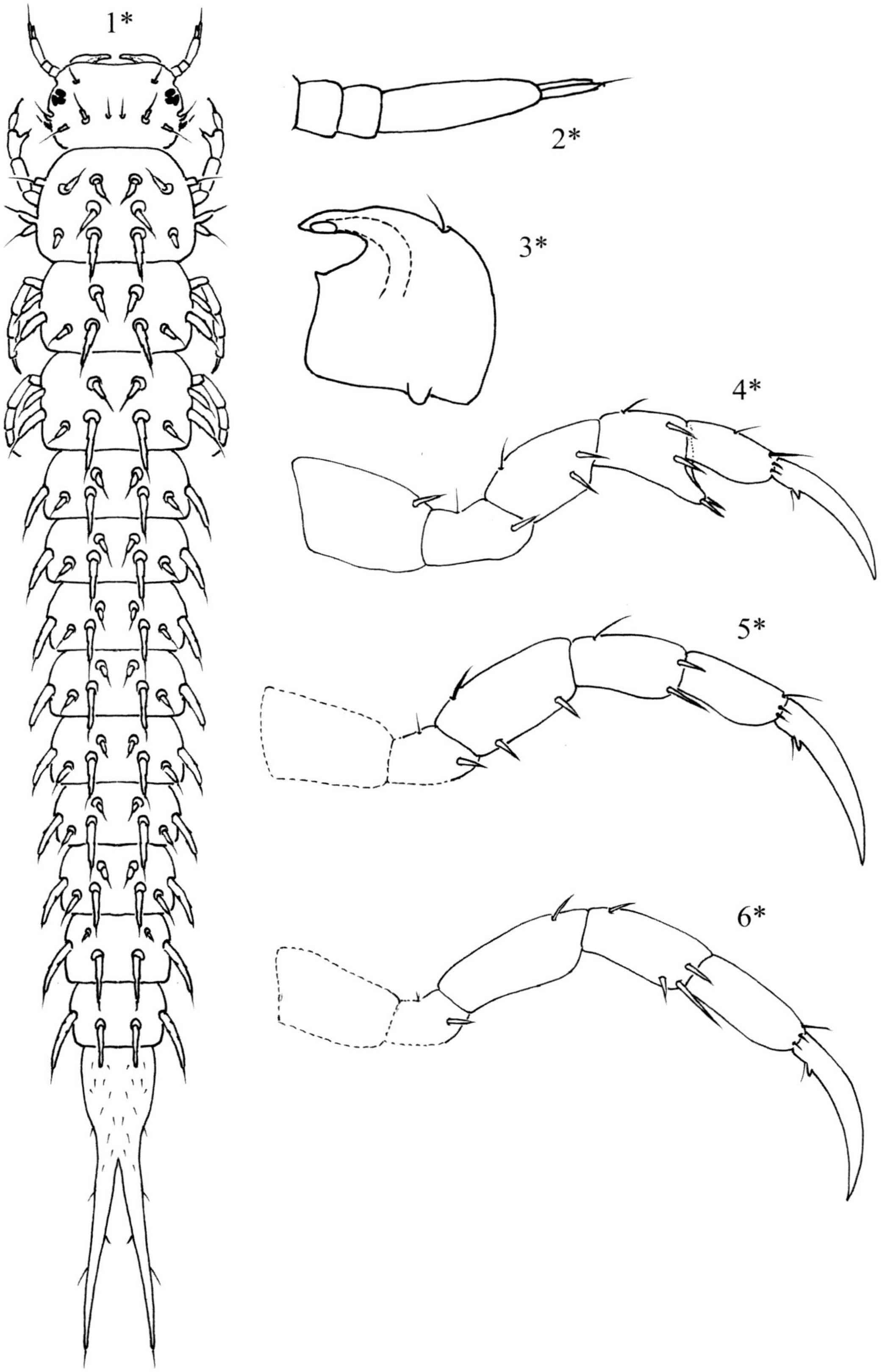

Plate 22. Figs. 1*–6*. *Haliplus (Haliplus) heydeni.* 1st instar larva. 1*, larva in dorsal view; 2*, antenna; 3*, mandible; 4*, foreleg; 5*, midleg; 6*, hindleg.

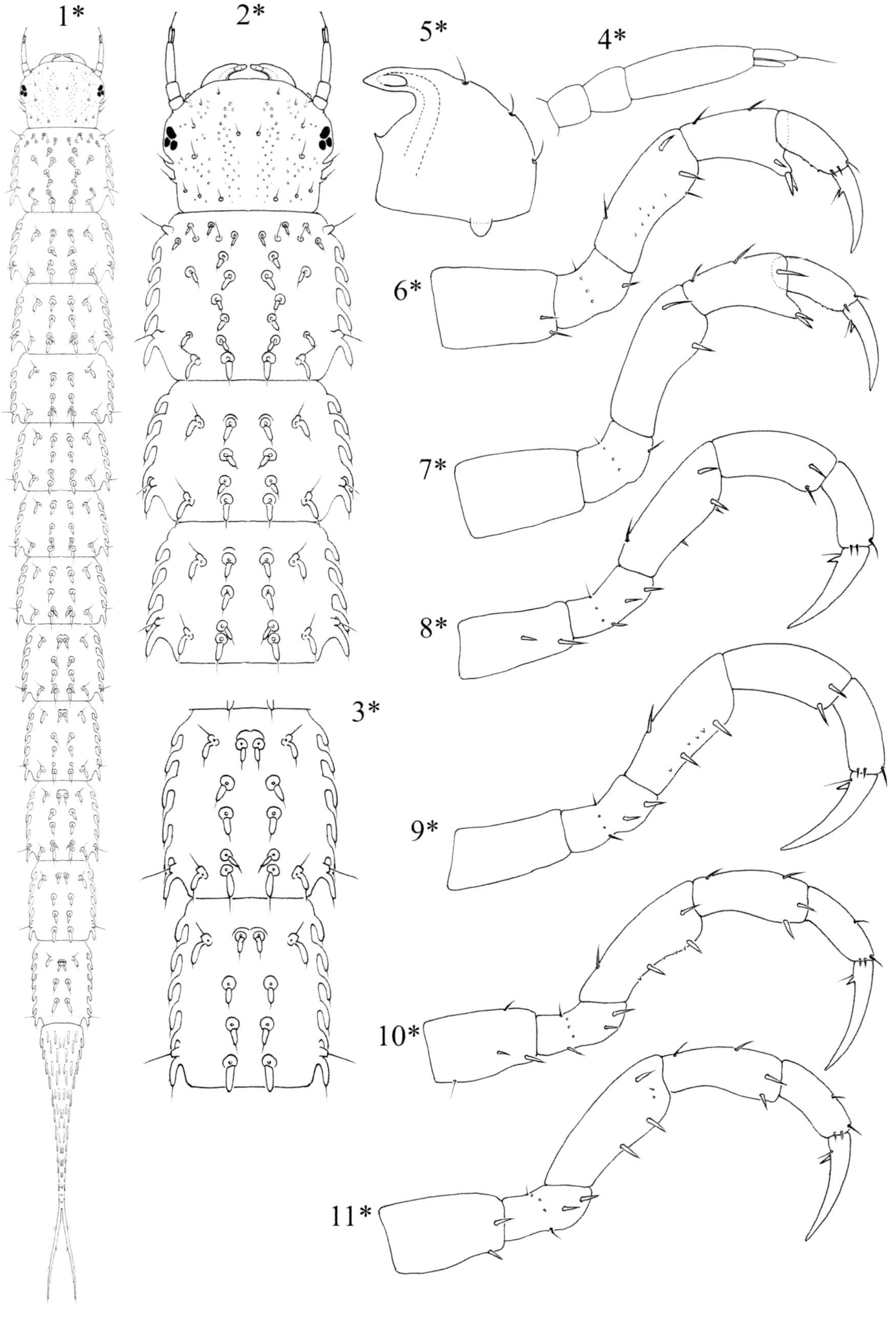

Plate 23. Figs. 1*–11*. *Haliplus (Haliplus) heydeni*. 2nd instar larva. 1*, body in dorsal view; 2*, head and thorax in dorsal view; 3*, abdominal segment VIII and IX in dorsal view; 4*, antenna; 5*, mandible; 6*, foreleg in dorsal view; 7*, foreleg in ventral view; 8*, midleg in dorsal view; 9*, midleg in ventral view; 10*, hindleg in dorsal view; 11*, hindleg in ventral view.

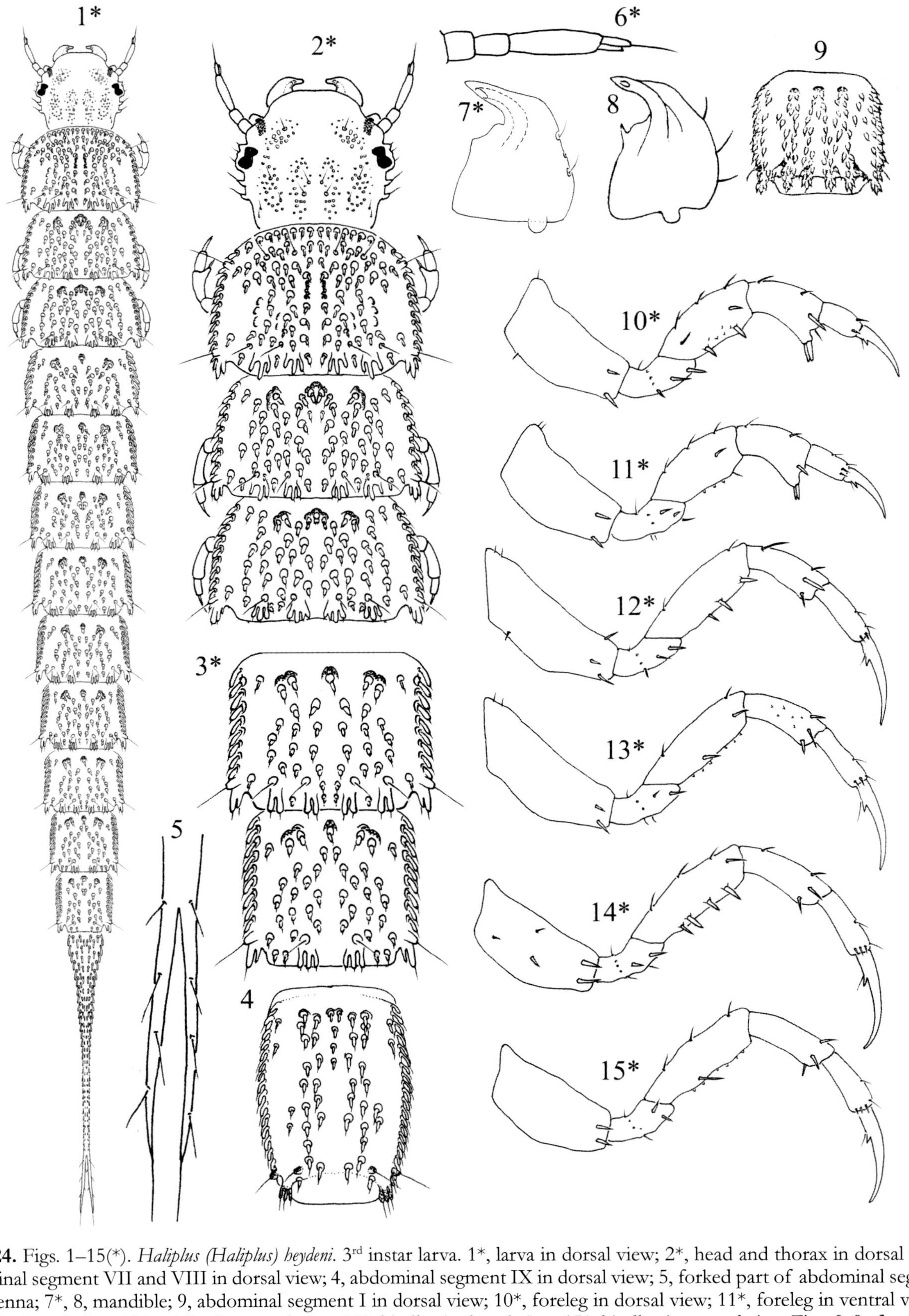

Plate 24. Figs. 1–15(*). *Haliplus (Haliplus) heydeni*. 3rd instar larva. 1*, larva in dorsal view; 2*, head and thorax in dorsal view; 3*, abdominal segment VII and VIII in dorsal view; 4, abdominal segment IX in dorsal view; 5, forked part of abdominal segment X; 6*, antenna; 7*, 8, mandible; 9, abdominal segment I in dorsal view; 10*, foreleg in dorsal view; 11*, foreleg in ventral view; 12*, midleg in dorsal view; 13*, midleg in ventral view; 14*, hindleg in dorsal view; 15*, hindleg in ventral view. Figs. 8, 9 after Bertrand (1942). Figs. 4, 5 after Vondel (1997).

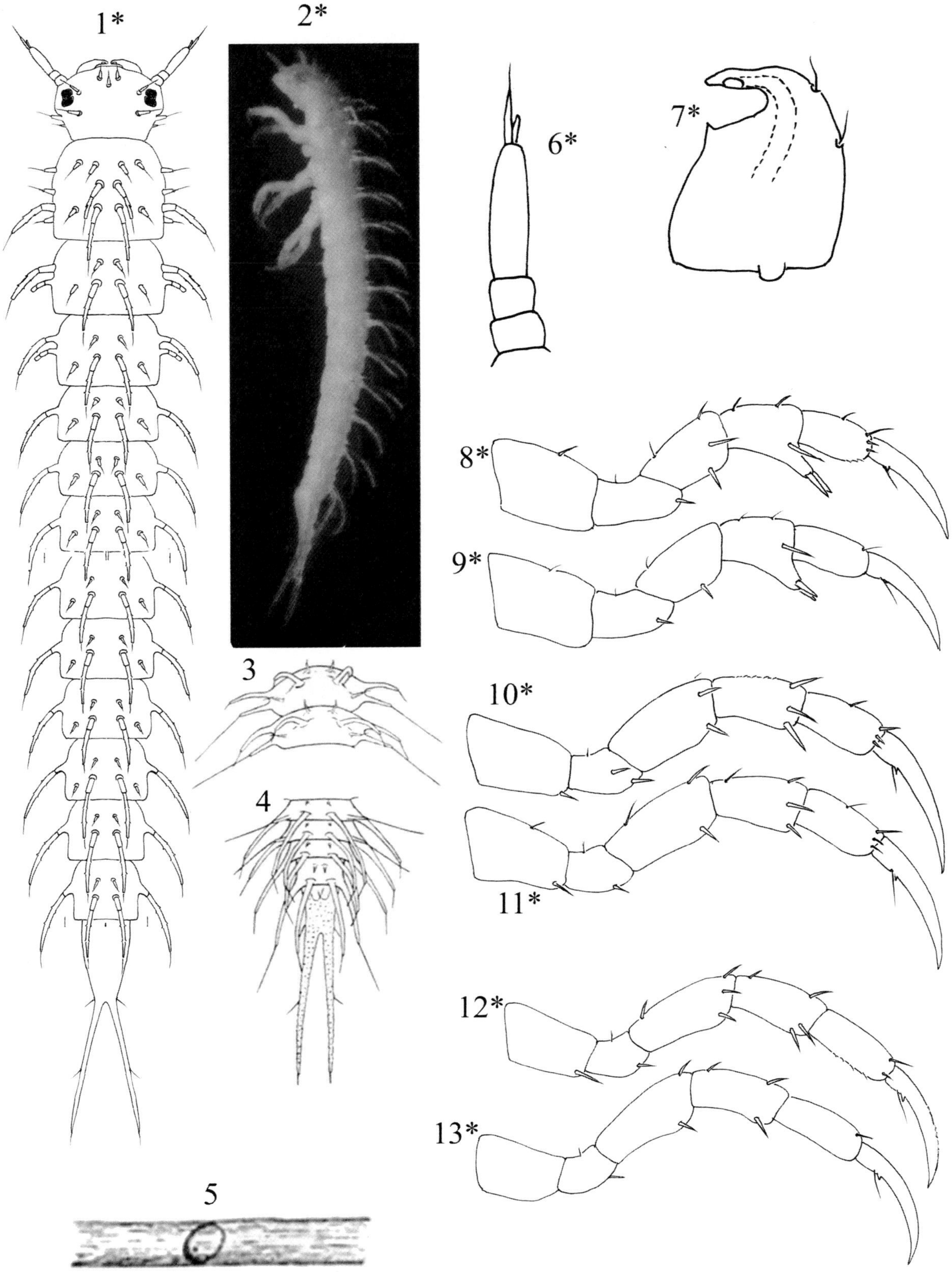

Plate 25. Figs. 1–13(*). *Haliplus (Haliplus) immaculatus.* 1st instar larva. 1*, body in dorsal view; 2*, larva in lateral view; 3, pro– and mesonotum; 4, last five abdominal segments in dorsal view; 5, egg; 6*, antenna; 7*, mandible; 8*, foreleg in dorsal view; 9*, foreleg in ventral view; 10*, midleg in dorsal view; 11*, midleg in ventral view; 12*, hindleg in dorsal view; 13*, hindleg in ventral view. Figs. 3–5 after Falkenström (1926).

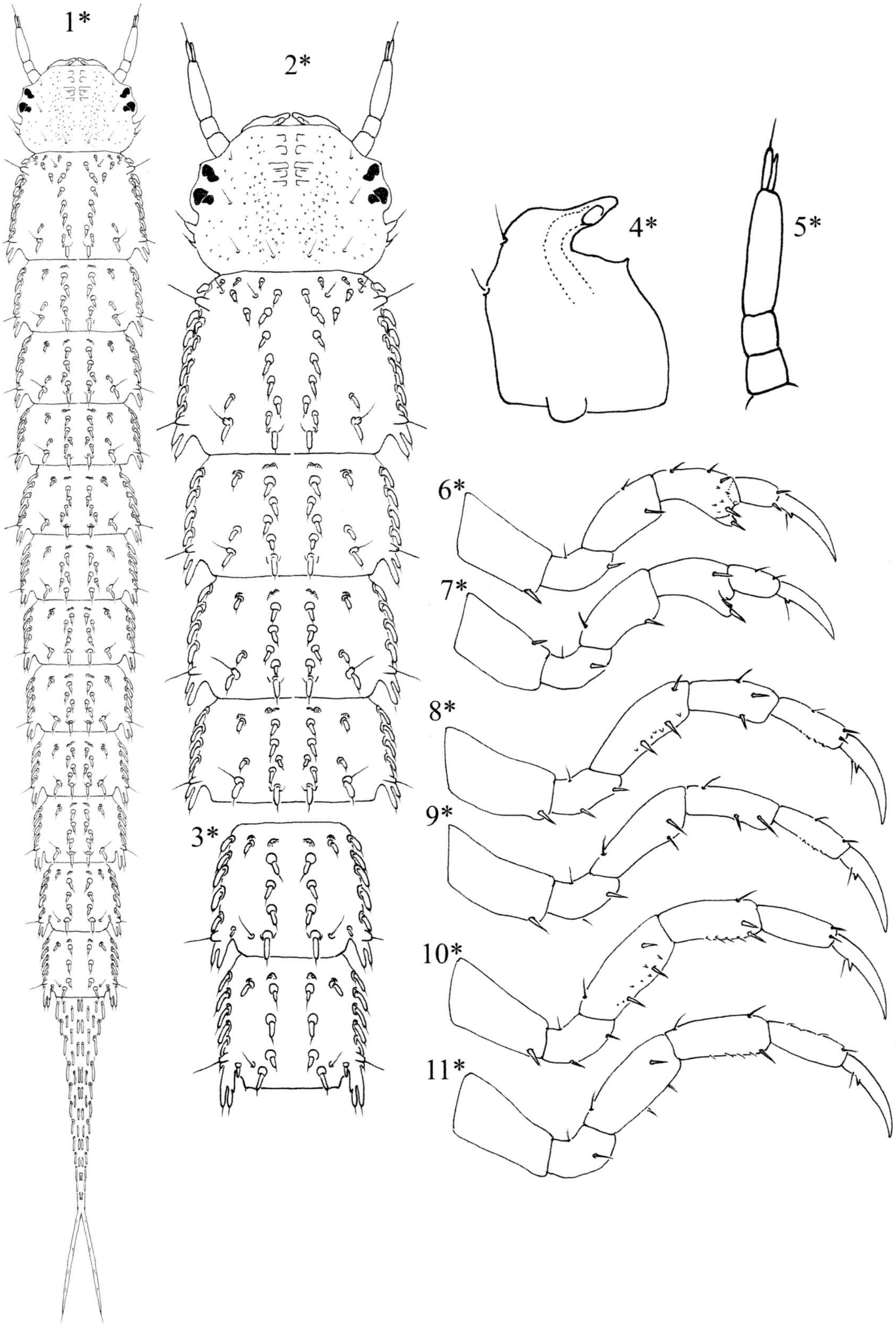

Plate 26. Figs. 1*–11*. *Haliplus (Haliplus) immaculatus*. 2nd instar larva. 1*, body in dorsal view; 2*, head, thorax and abdominal segment I in dorsal view; 3*, abdominal segment VIII and IX in dorsal view; 4*, mandible'5*, antenna; 6*, foreleg in dorsal view; 7*, foreleg in ventral view; 8*, midleg in dorsal view; 9*, midleg in ventral view; 10*, hindleg in dorsal view; 11*, hindleg in ventral view.

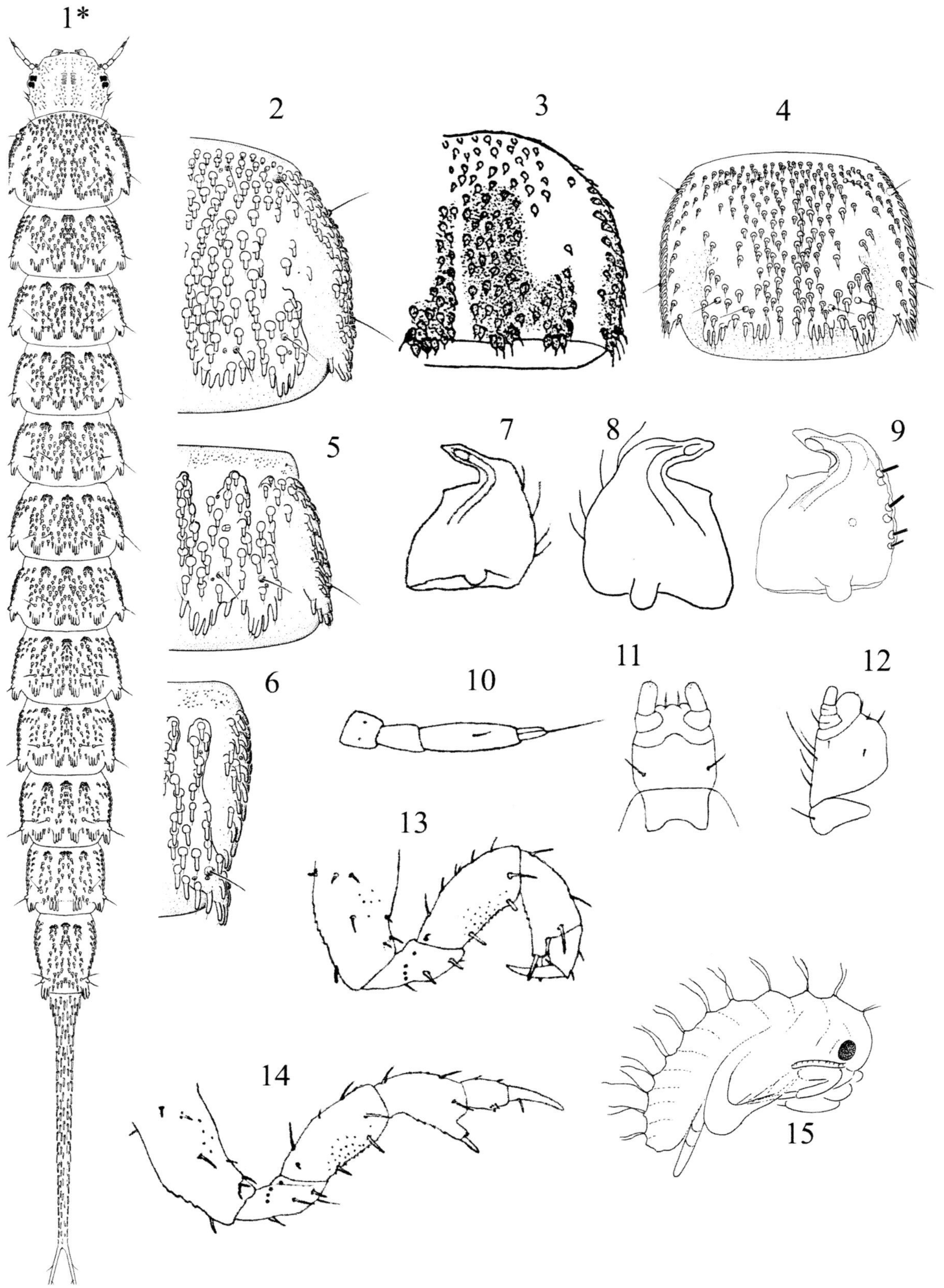

Plate 27. Figs. 1*–15. *Haliplus (Haliplus) immaculatus*. 3rd instar larva and pupa. 1*, body in dorsal view; 2–4, prosternum; 5, abdominal segment I in dorsal view; 6, abdominal segment IX in dorsal view; 7–9, mandible; 10, antenna; 11, labium; 12, maxilla; 13, 14, foreleg; 15, pupa. Figs. 7, 10, 11, 12, 13, 14 after Falkenström (1926). Figs. 3, 8 after Bertrand (1950). Figs. 2, 5, 6, 9 after Seeger (1971a). Figs. 4, 15 after Vondel (1997).

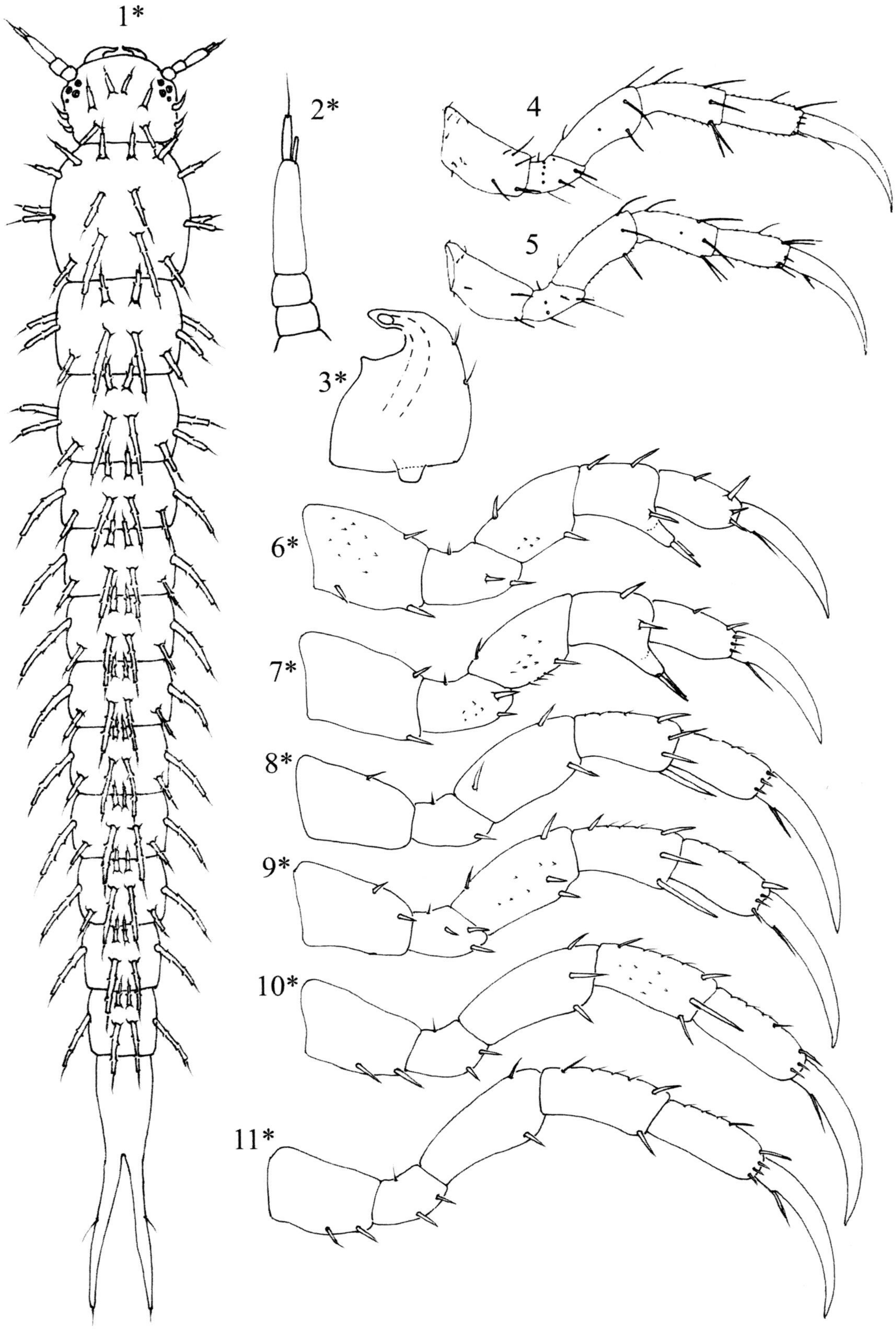

Plate 28. Figs. 1–11(*). *Haliplus (Haliplus) lineolatus.* 1st instar larva. 1*, body in dorsal view; 2*, antenna; 3*, mandible; 4, 10*, hindleg in dorsal view; 5, 11*, hindleg in ventral view; 6*, foreleg in dorsal view; 7*, foreleg in ventral view; 8*, midleg in dorsal view; 9*, midleg in ventral view. Figs. 4, 5 after Nilsson (1988).

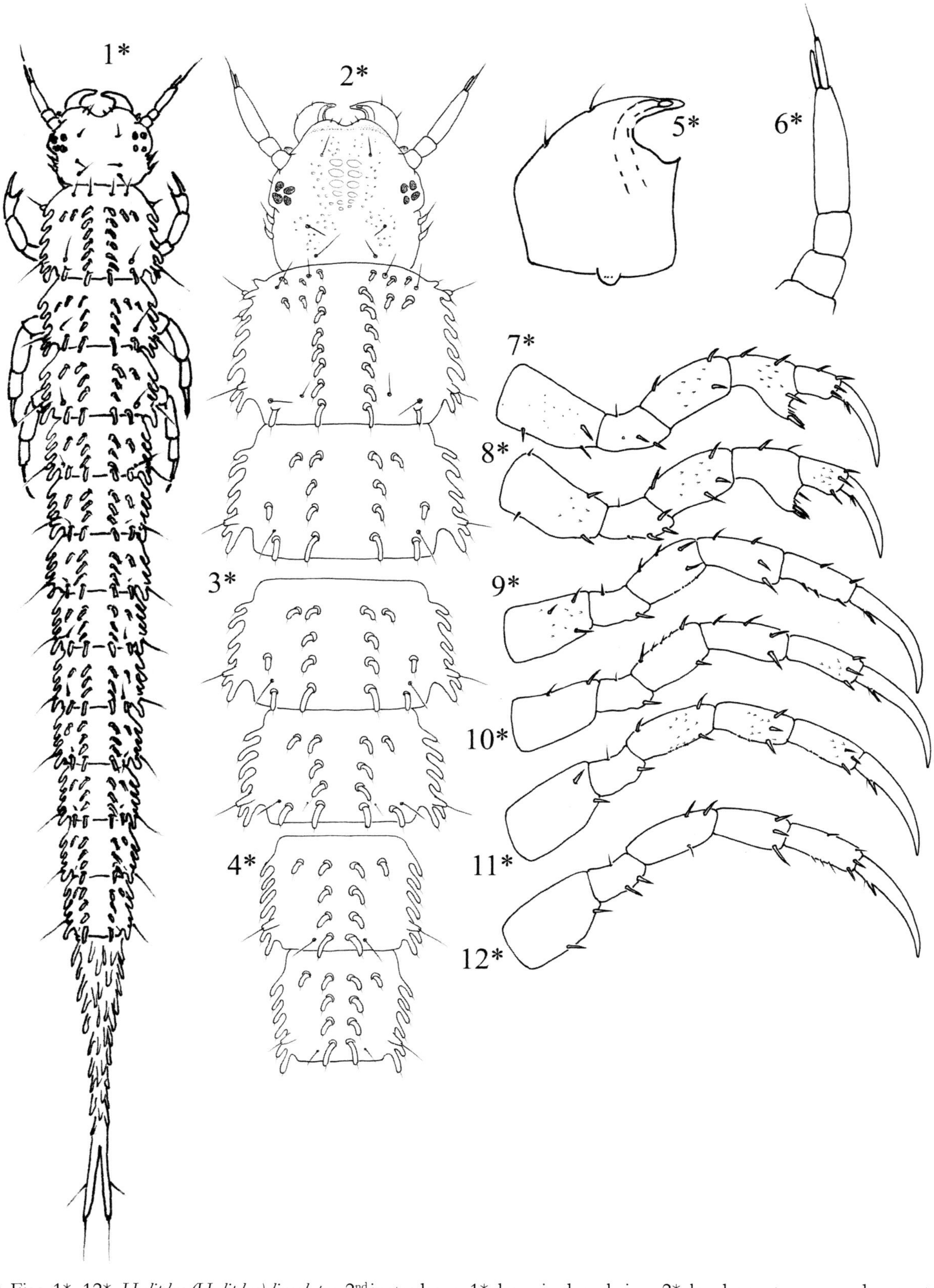

Plate 29. Figs. 1*–12*. *Haliplus (Haliplus) lineolatus*. 2nd instar larva. 1*, larva in dorsal view; 2*, head, prosternum and mesosternum; 3*, abdominal segment I and II in dorsal view; 4*, abdominal segment VIII and IX in dorsal view; 5*, mandible; 6*, antenna; 7*, foreleg in dorsal view; 8*,foreleg in ventral view; 9*, midleg in dorsal view; 10*, midleg in ventral view; 11*, hindleg in dorsal view; 12*, hindleg in ventral view.

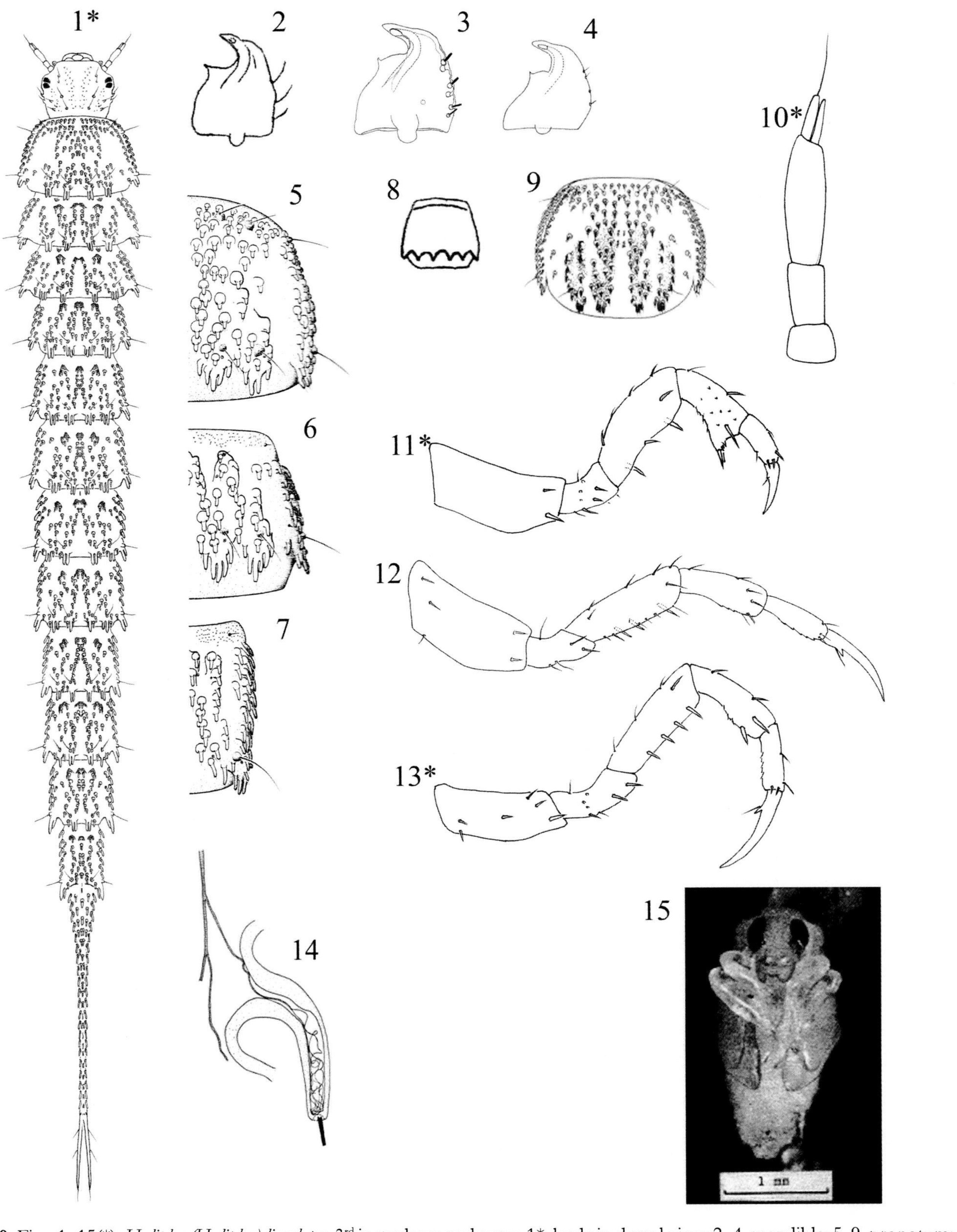

Plate 30. Figs. 1–15(*). *Haliplus (Haliplus) lineolatus*. 3rd instar larva and pupa. 1*, body in dorsal view; 2–4, mandible; 5, 9, pronotum; 6, 8, abdominal segment I in dorsal view; 7, abdominal segment IX in dorsal view; 10*, antenna; 11*, foreleg; 12, midleg; 13*, hindleg; 14, tubercle on dorsal and lateral side of the body in cross–section. Figs. 2, 8 after Bertrand (1931). Figs. 3, 5–7, 14, 15 after Seeger (1971a). Figs. 4, 9, 12 after Vondel (1997).

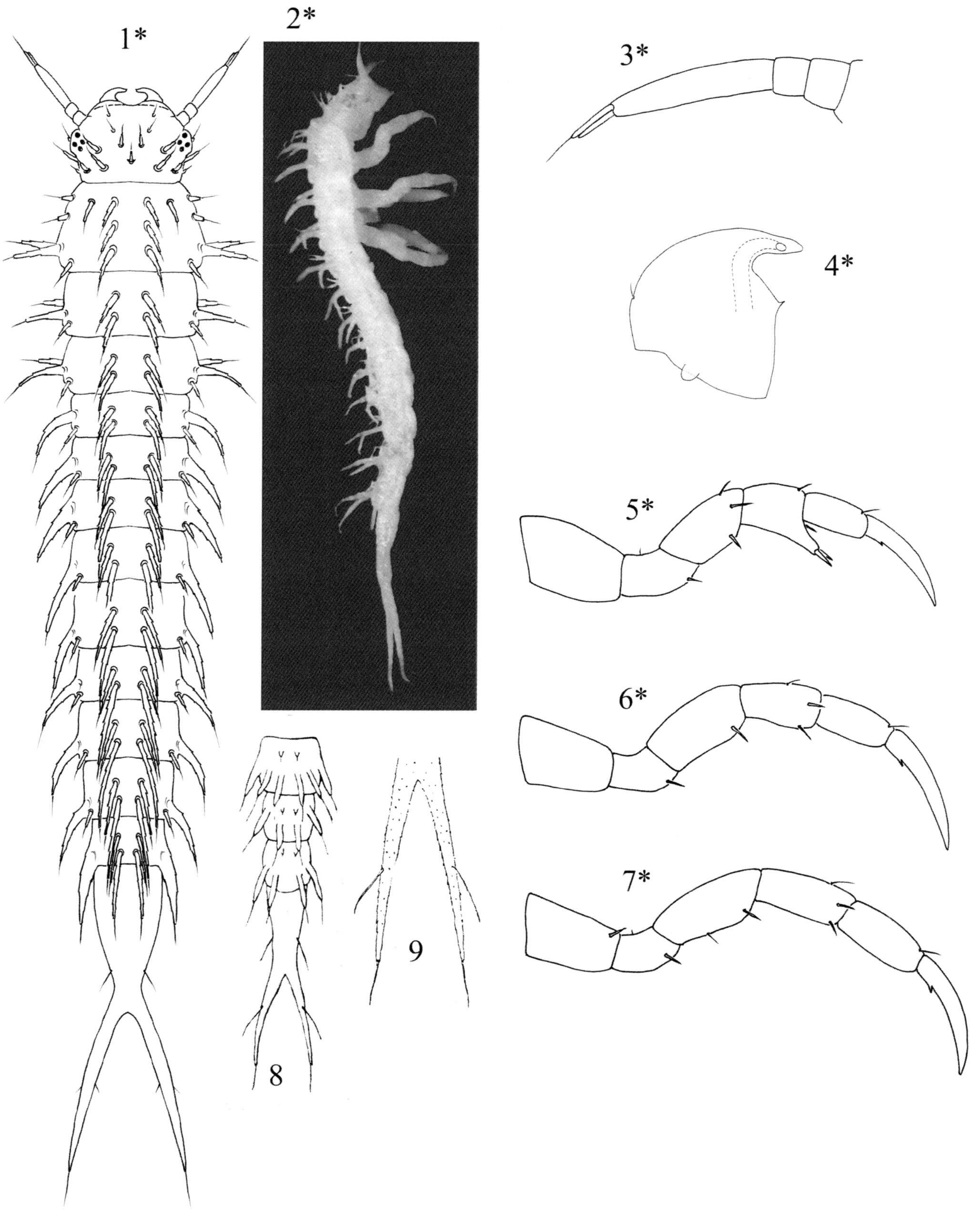

Plate 31. Figs. 1–9 (*). *Haliplus (Haliplus) ruficollis*. 1st instar larva. 1*, body in dorsal view; 2*, larva in lateral view; 3*, antenna; 4*, mandible; 5*, foreleg; 6*, midleg; 7*, hindleg; 8, last four abdominal segments in dorsal view; 9, forked part of abdominal segment X. Figs. 8, 9 after Bertrand (1923).

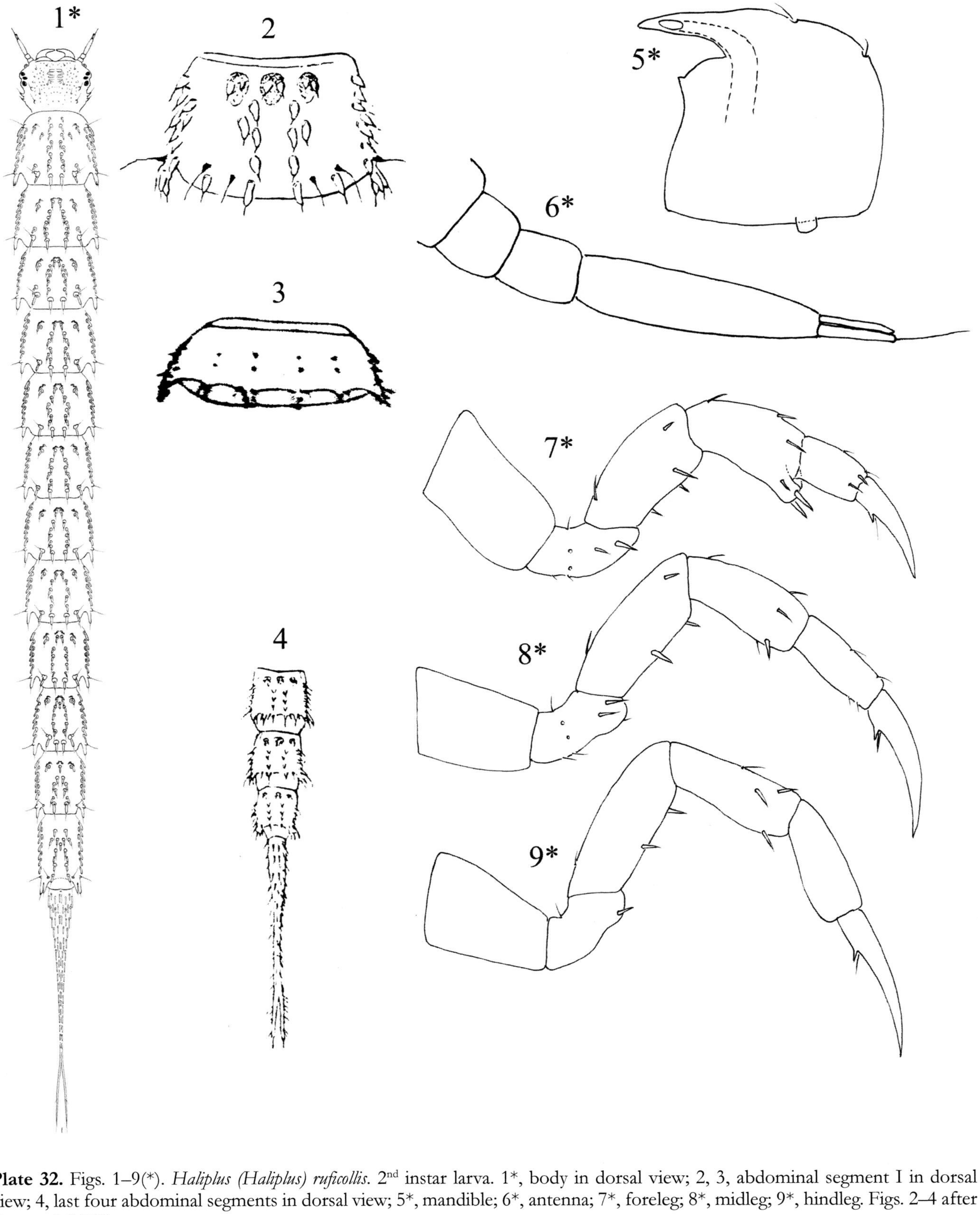

Plate 32. Figs. 1–9(*). *Haliplus (Haliplus) ruficollis*. 2nd instar larva. 1*, body in dorsal view; 2, 3, abdominal segment I in dorsal view; 4, last four abdominal segments in dorsal view; 5*, mandible; 6*, antenna; 7*, foreleg; 8*, midleg; 9*, hindleg. Figs. 2–4 after Bertrand (1923).

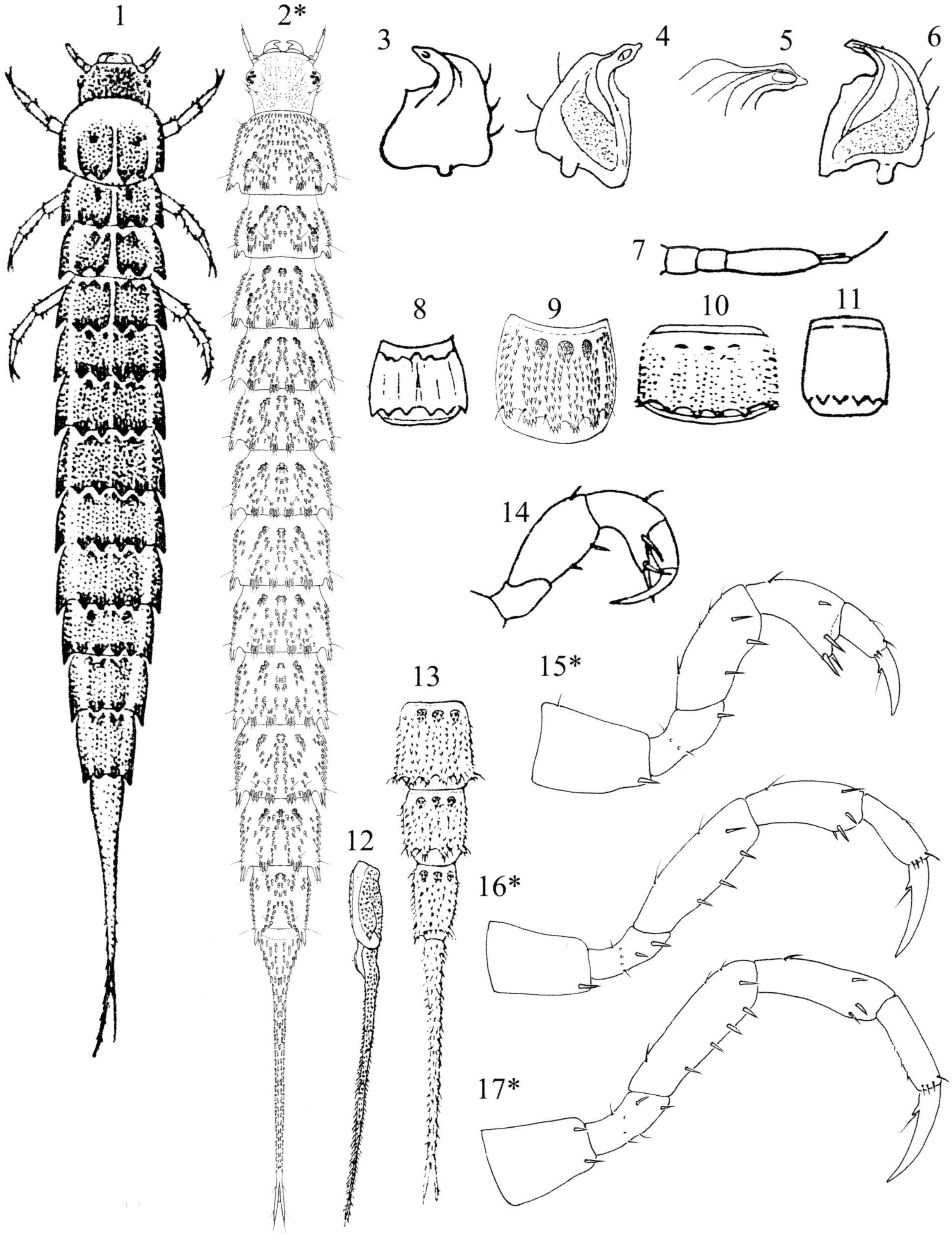

Plate 33. Figs. 1–17(*). *Haliplus (Haliplus) ruficollis.* 3rd instar larva. 1, 1*, larva in dorsal view; 3, 4, 5, 6, mandible; 7, antenna; 8–11, abdominal segment I in dorsal view; 12, last two abdominal segments in lateral view; 13, last four abdominal segments in dorsal view; 14, 15*, foreleg; 16*, midleg; 17*, hindleg. Fig. 6 after Schiødte (1864). Figs. 1, 14 after Rousseau (1919b). Figs. 4, 5, 7, 8, 9, 12, 13 after Bertrand (1923). Fig. 3, 10 after Bertrand (1928b). Fig. 11 after Bertrand (1931).

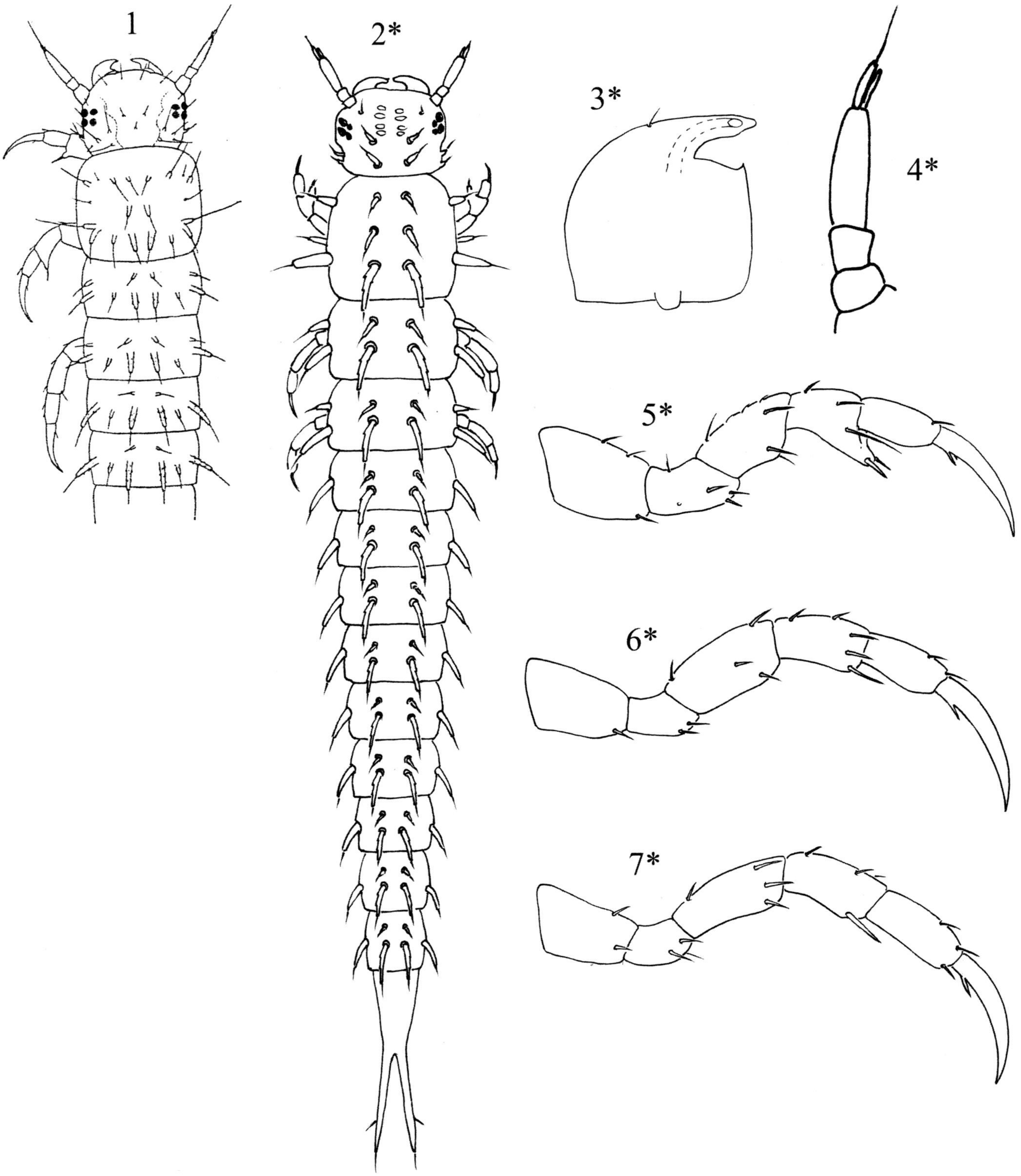

Plate 34. Figs. 1–7(*). *Haliplus (Haliplus) sibiricus.* 1st instar larva. 1, head, thorax and first two abdominal segments in dorsal view; 2* larva in dorsal view; 3*, mandible; 4*, antenna; 5*, foreleg; 6*, midleg; 7*, hindleg. Fig. 1 after Beier (1929), but see discussion.

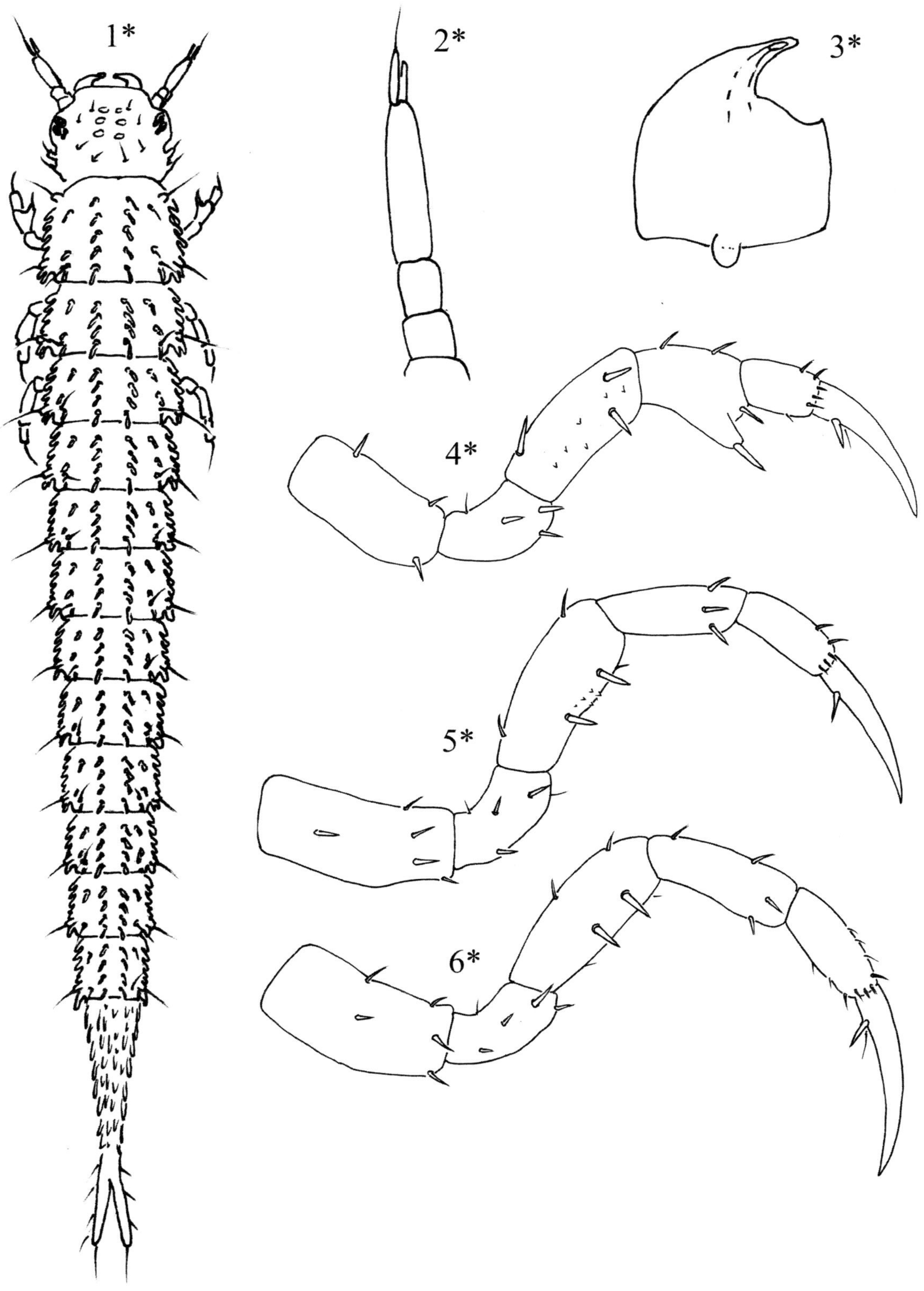

Plate 35. Figs. 1*–6*. *Haliplus (Haliplus) sibiricus*. 2nd instar larva. 1*, larva in dorsal view; 2*, antenna; 3*, mandible; 4*, foreleg; 5*. midleg; 6*, hindleg.

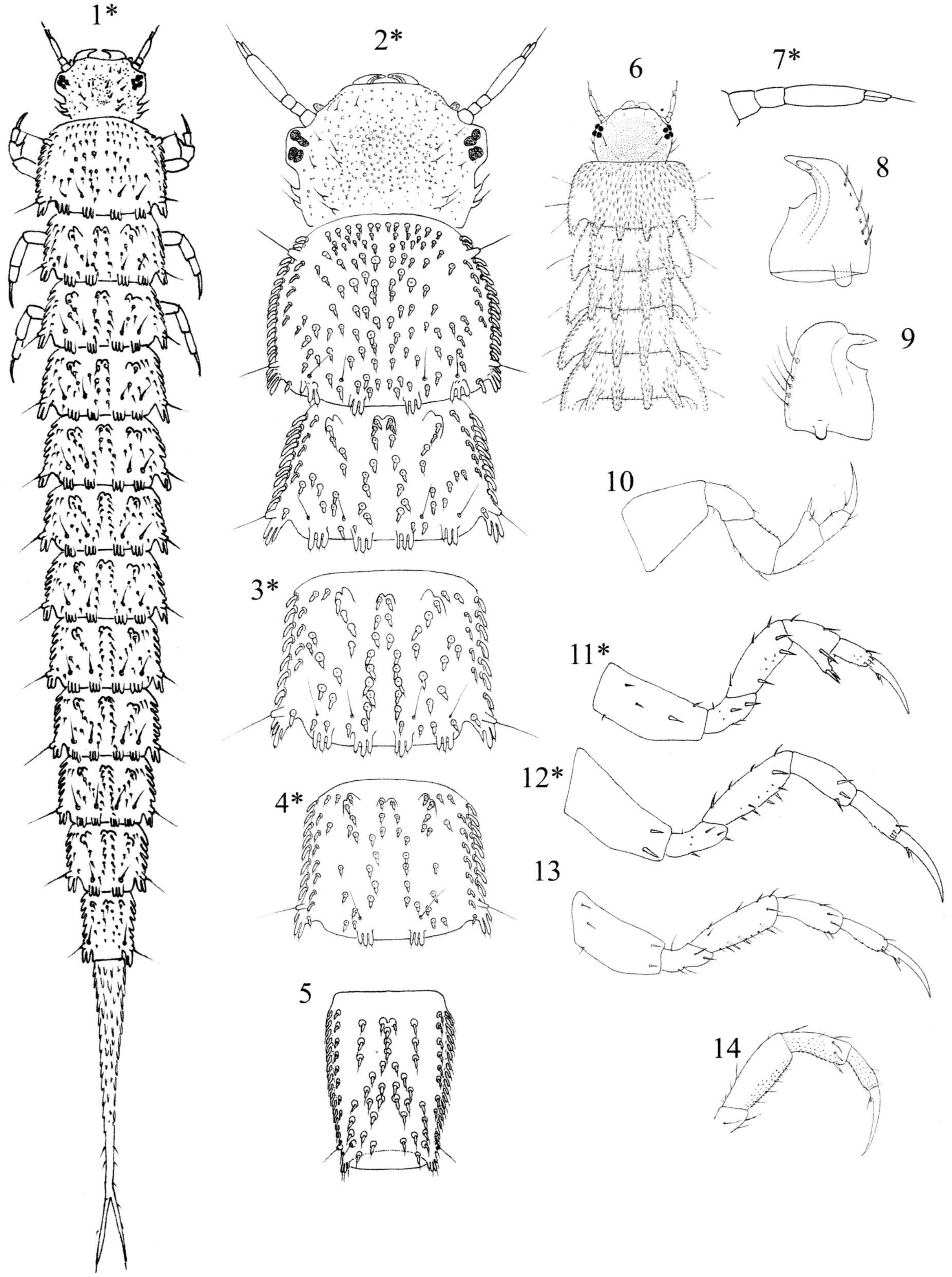

Plate 36. Figs. 1–14(*). *Haliplus (Haliplus) sibiricus.* 3rd instar larva. 1*, larva in dorsal view; 2*, head, pronotum and mesonotum; 3*, abdominal segment I in dorsal view; 4*, abdominal segment VIII in dorsal view; 5, abdominal segment IX in dorsal view; 6, head, thorax and two first abdominal segments in dorsal view; 7*, antenna; 8, 9, mandible; 10, 11*, foreleg; 12*, midleg; 13, 14, hindleg. Figs. 6, 9, 10, 14 after Beier (1929), but see discussion. Figs. 5, 8, 13 after Vondel (1997).

Plate 37. Figs. 1–5. *Haliplus* sp. 1, *Haliplus (Haliplus) zacharenkoi*, 3rd instar larva; 2, *Haliplus (Liaphlus) angustifrons*, 3rd instar larva; 3, *Haliplus (Liaphlus) basinotatus*, 3rd instar larva; 4, 5, *Haliplus (Liaphlus) eximius*: 4, 3rd instar larva, 5, pupa. Fig. 1 after Dyaditchko (2009). Fig. 2 after Yamasaki (2024). Fig. 3 after Mitamura (2017). Figs. 4, 5 after Yamazaki et al. (2020).

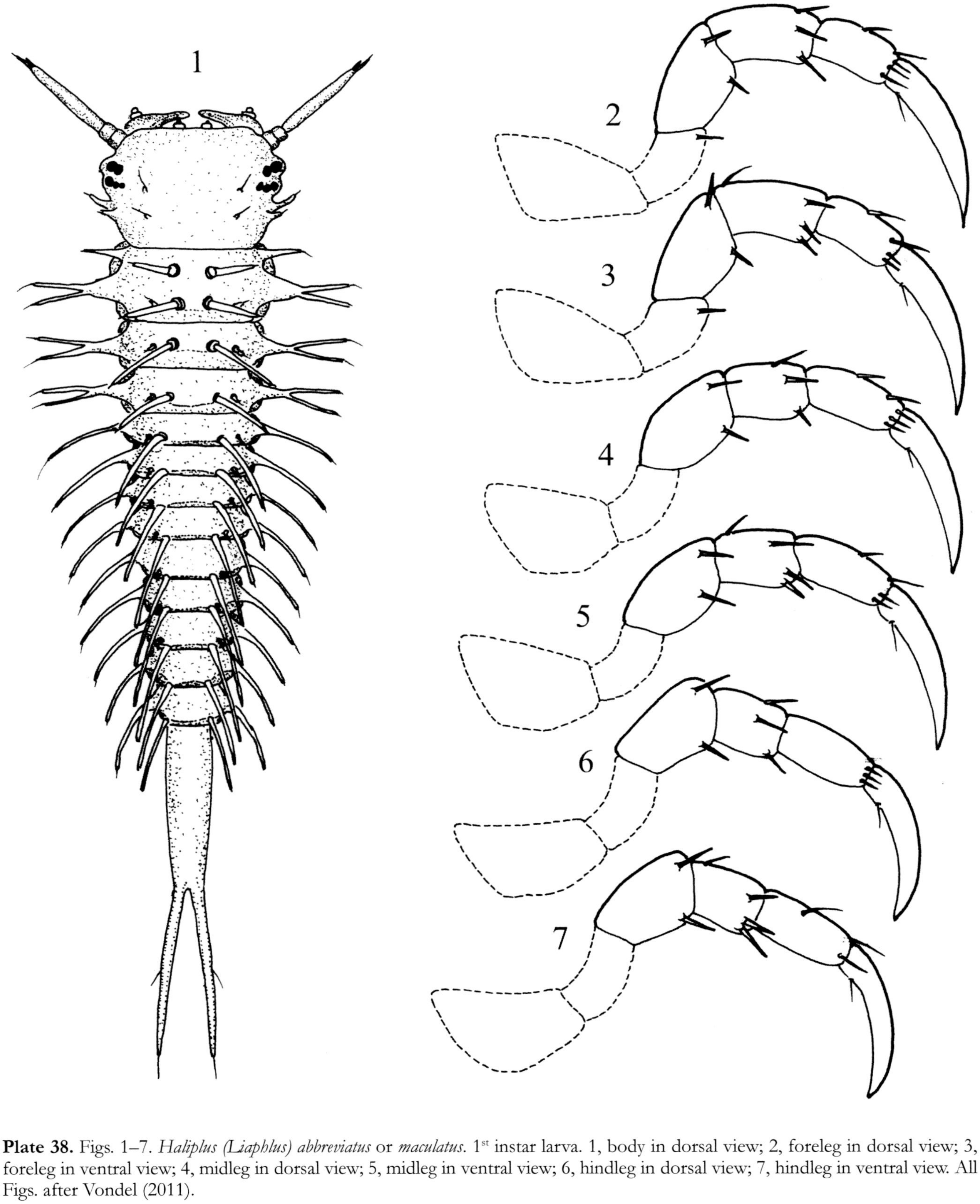

Plate 38. Figs. 1–7. *Haliplus (Liaphlus) abbreviatus* or *maculatus*. 1st instar larva. 1, body in dorsal view; 2, foreleg in dorsal view; 3, foreleg in ventral view; 4, midleg in dorsal view; 5, midleg in ventral view; 6, hindleg in dorsal view; 7, hindleg in ventral view. All Figs. after Vondel (2011).

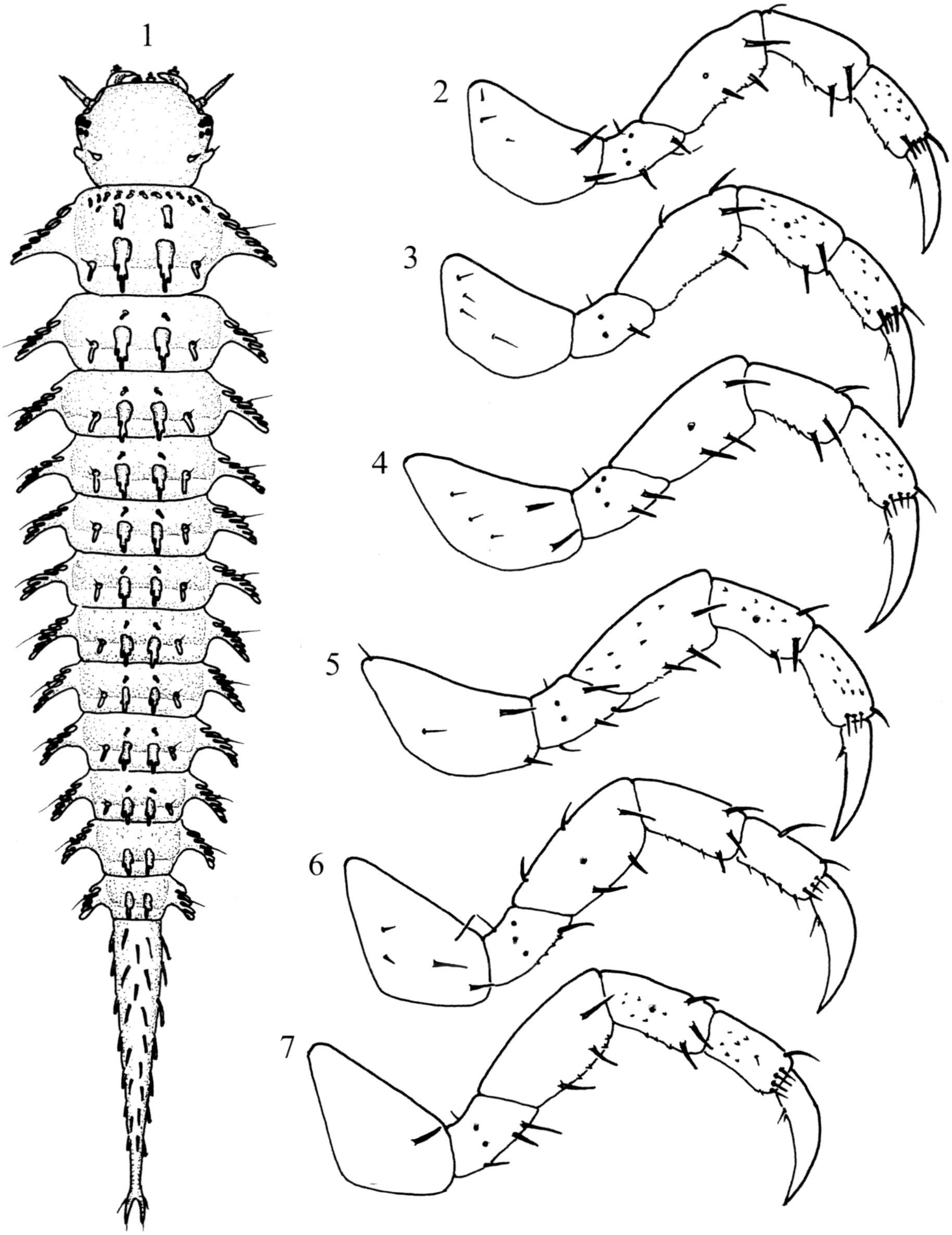

Plate 39. Figs. 1–7. *Haliplus (Liaphlus) abbreviatus* or *maculatus*. 2nd instar larva. 1, body in dorsal view; 2, foreleg in dorsal view; 3, foreleg in ventral view; 4, midleg in dorsal view; 5, midleg in ventral view; 6, hindleg in dorsal view; 7, hindleg in ventral view. All Figs. after Vondel (2011).

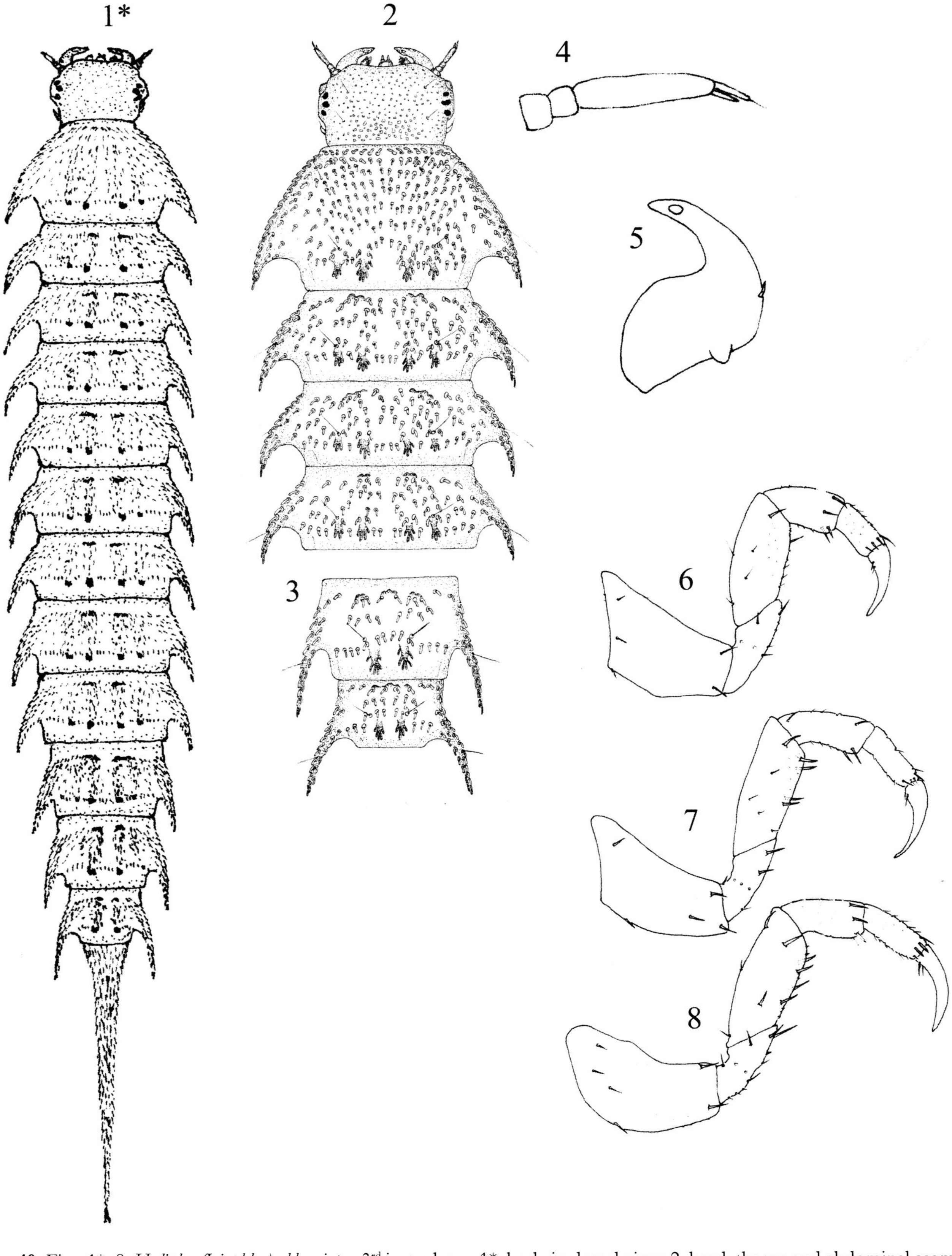

Plate 40. Figs. 1*–8. *Haliplus (Liaphlus) abbreviatus*. 3rd instar larva. 1*, body in dorsal view; 2, head, thorax and abdominal segment I in dorsal view; 3, abdominal segment VIII and IX in dorsal view; 4, antenna; 5, mandible; 6, foreleg; 7, midleg; 8, hindleg. All Figs. after Vondel (2011), but 1* modified.

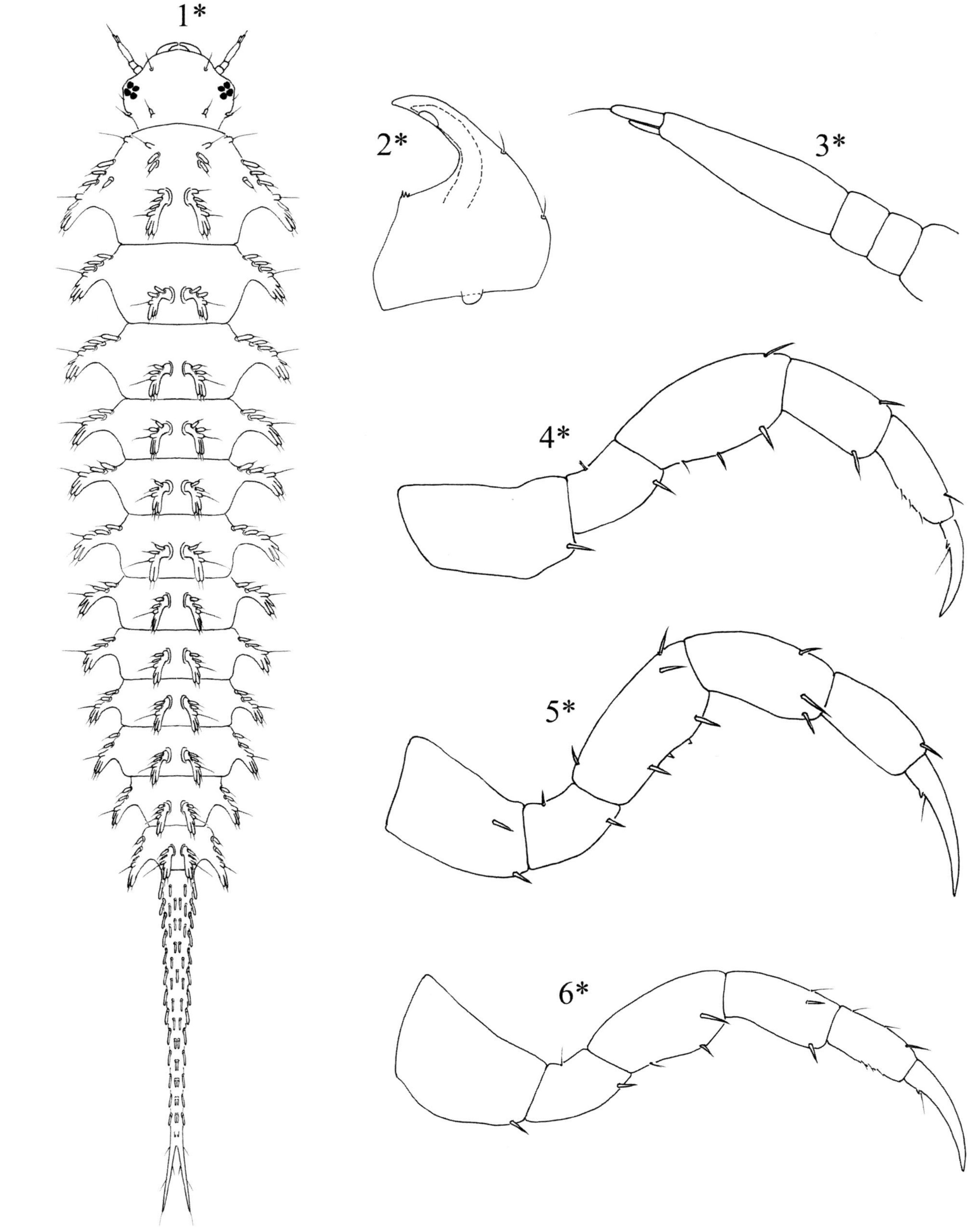

Plate 41. Figs. 1*–6*. *Haliplus (Liaphlus) flavicollis*. 2nd instar larva. 1*, body in dorsal view; 2*, mandible; 3*, antenna; 4*, foreleg; 5*, midleg, 6*, hindleg.

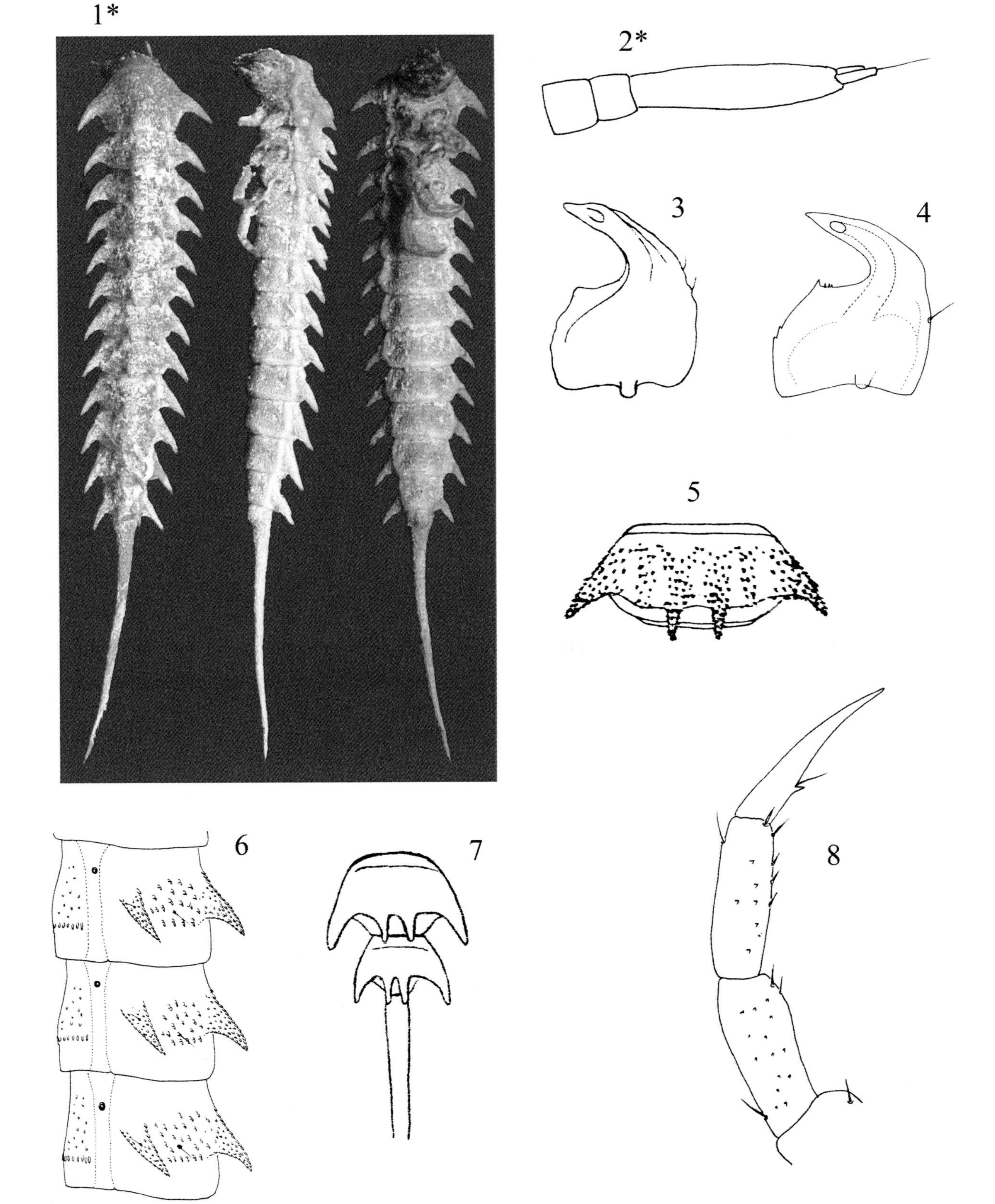

Plate 42. Figs. 1*–8. *Haliplus (Liaphlus) flavicollis.* 3rd instar larva. 1*, larva in dorsal, lateral and ventral view; 2*, antenna; 3, 4, mandible; 5, abdominal segment I in dorsal view; 6, abdominal segments in lateral view; 7, abdominal segment VIII and IX in dorsal view; 8, last three segments of hindleg. Figs. 3, 5, 7 after Bertrand (1928b). Figs. 4, 6, 8 after Vondel (1997).

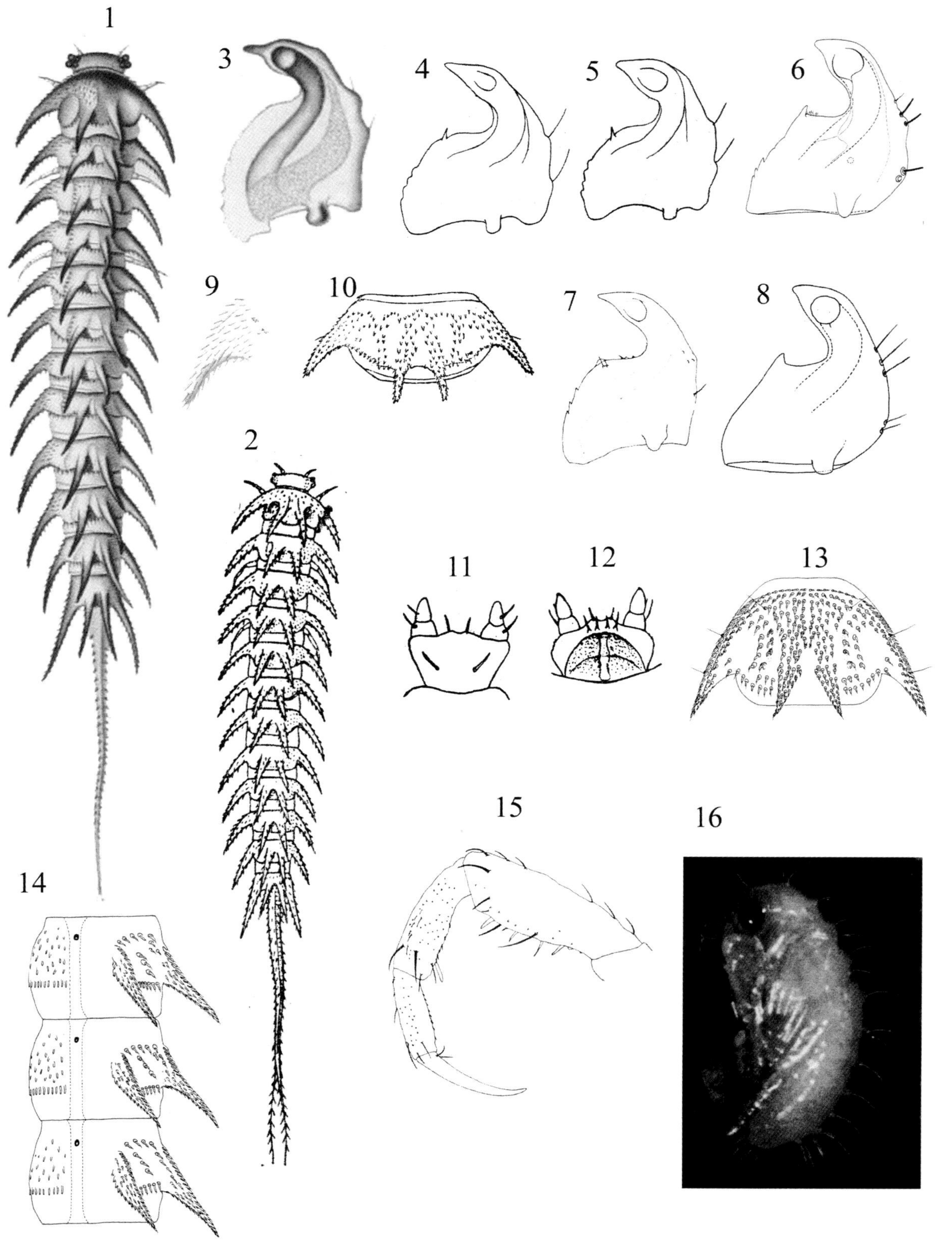

Plate 43. Figs. 1–16. *Haliplus (Liaphlus) fulvus.* 3rd instar larva and pupa. 1, 2, larva in dorsal view; 3–8, mandible; 9, lateral process of thoracic and abdominal segments; 10, abdominal segment I in dorsal view; 11, labium in ventral view; 12, labium in dorsal view; 13, pronotum; 14, abdominal segments in lateral view; 15, foreleg; 16, pupa. Figs. 1, 3, 9 after Schiødte (1864). Fig. 2 after Roussaeu (1919b). Figs. 4, 5, 10, 11, 12 after Bertrand (1928b). Figs. 6, 16 after Seeger (1971), but see discussion. Figs. 7, 15 after Holmen (1981). Figs. 8, 13, 14 after Vondel (1997).

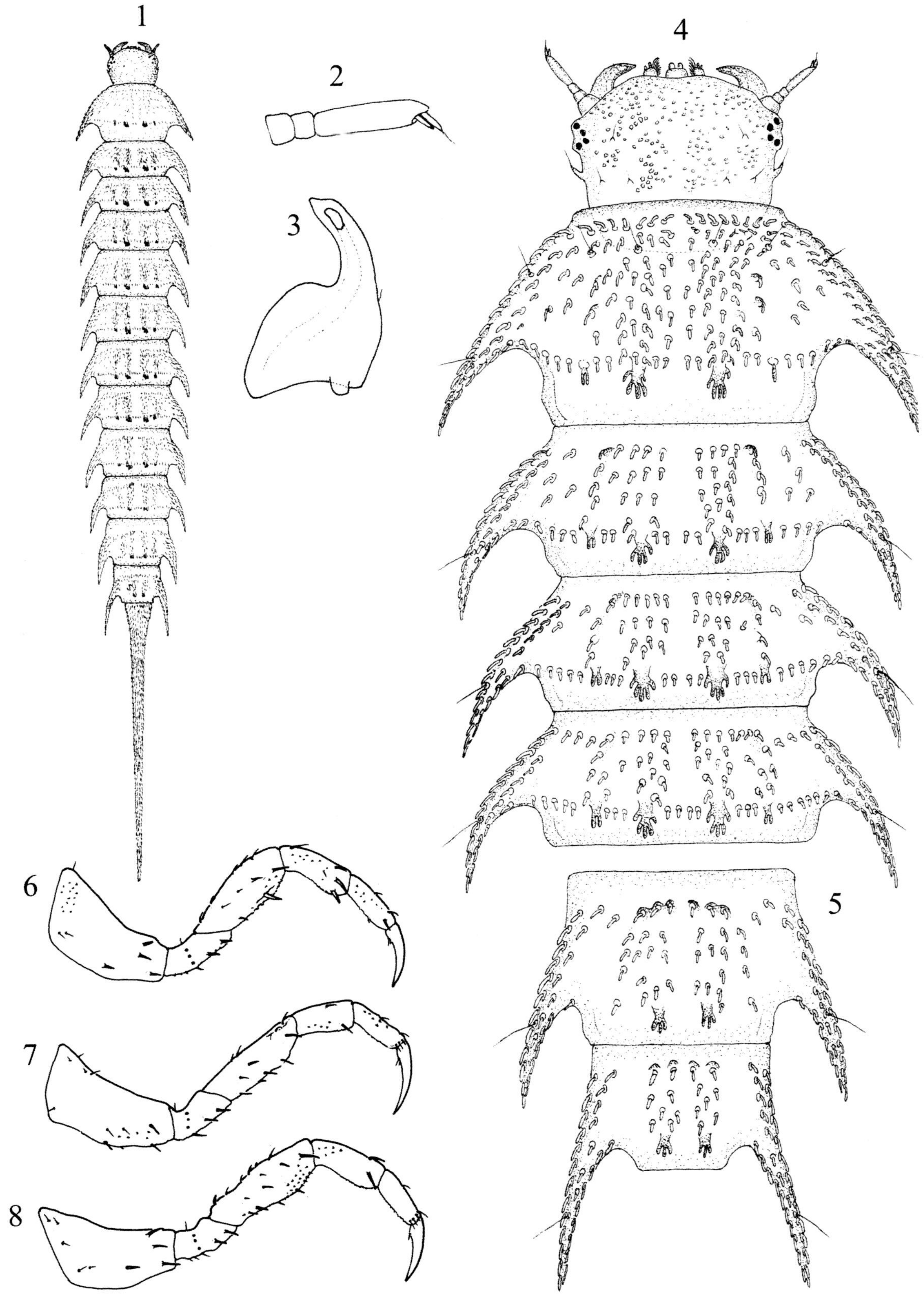

Plate 44. Figs. 1–8. *Haliplus (Liaphlus) maculatus.* 3rd instar larva. 1, body in dorsal view; 2, antenna; 3, mandible; 4, head, thorax and abdominal segment I in dorsal view; 5, eighth and ninth abdominal segment VIII and IX in dorsal view; 6, foreleg; 7, midleg; 8, hindleg. All Figs. after Vondel (2011).

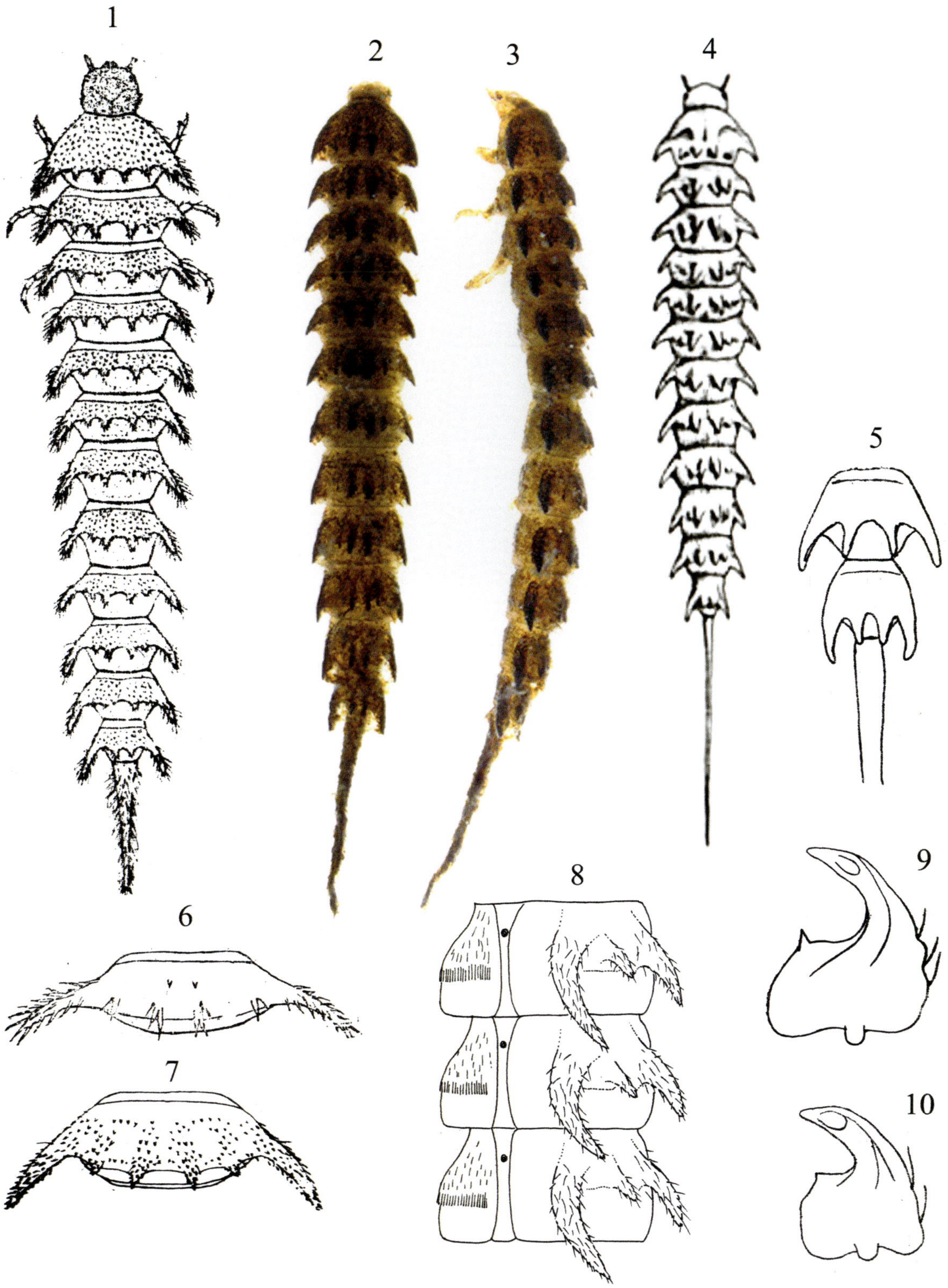

Plate 45. Figs. 1–10. *Haliplus* sp. Figs. 1, 6–9, *Haliplus (Liaphlus) mucronatus*. 1, 3rd instar larva in dorsal view; 6, abdominal segment I of 2nd instar in dorsal view; 7, abdominal segment I of 3rd instar in dorsal view; 8, abdominal segments of 3rd instar in lateral view; 9, mandible of 3rd instar. Figs. 2–4, *Haliplus (Liaphlus) ovalis*, 3rd instar larva, 2, 4 in dorsal view, 3 in lateral view. Figs. 5, 10, *Haliplus (Liaphlus) rubidus*, 3rd instar larva, 5, abdominal segment VIII and IX in dorsal view; 10, mandible. Figs. 1, 6, 7, 9 after Bertrand (1928b). Figs. 5, 10 after Bertrand (1931). Fig. 8 after Vondel (1997). Figs. 2, 3 after Hayashi (2011). Fig. 4 after Sato & Yoshitomi (2018).

Plate 46. Figs. 1–6. *Haliplus (Liaphlus) sharpi*. 1, 2nd instar larva in dorsal view; 2, 3, 3rd instar larva in dorsal view; 4–6, pupa in ventral, lateral and dorsal view. Figs. 1, 2, 4–6 after Nakanishi (2012). Fig. 3 after Mitamura (2017).

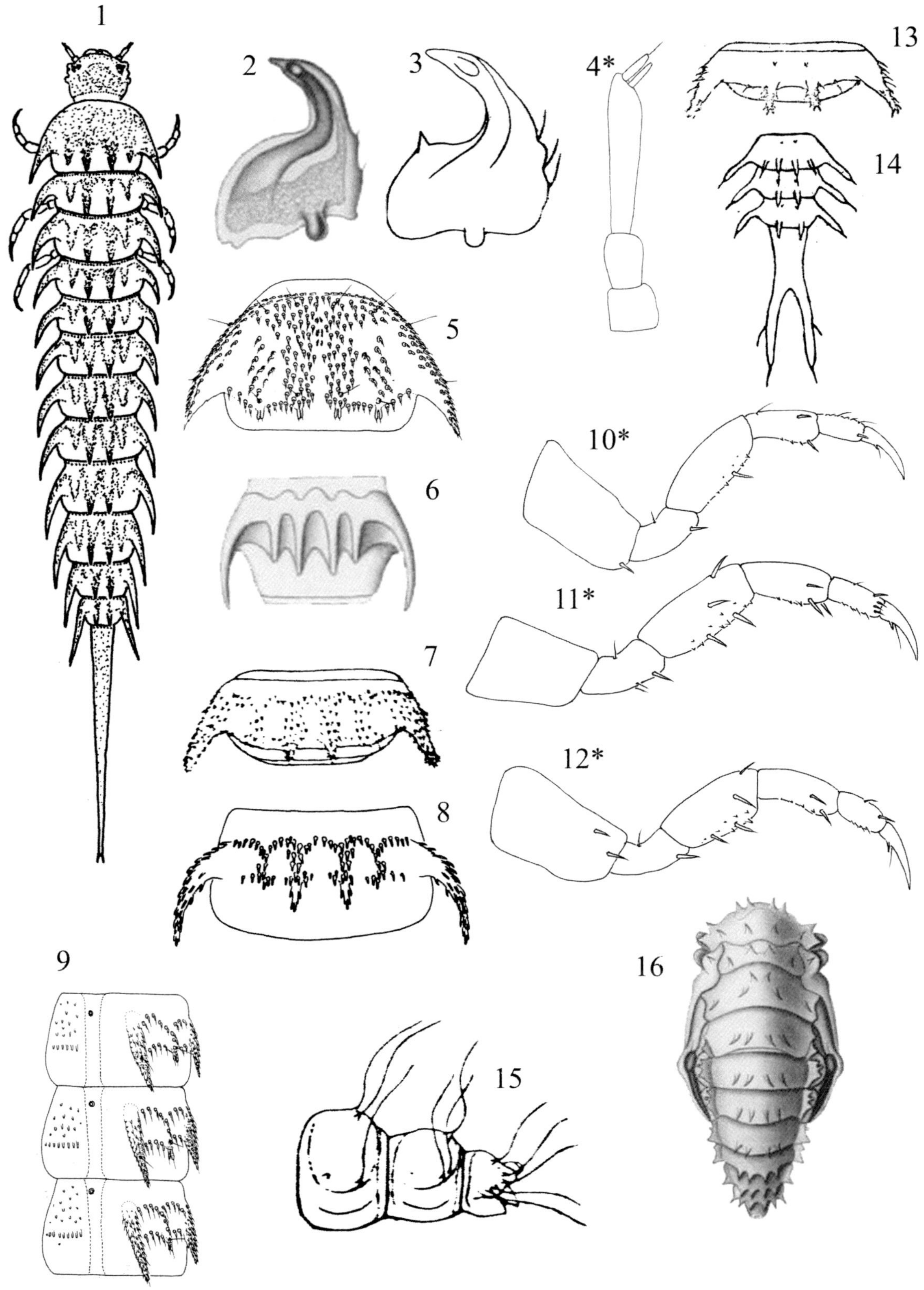

Plate 47. Figs. 1–16(*). *Haliplus (Liaphlus) variegatus*. Figs. 1–12(*). 3rd instar larva: 1, larva in dorsal view; 2, 3, mandible; 4*. antenna; 5, pronotum; 6–8, abdominal segment I in dorsal view; 9, abdominal segments in lateral view; 10*, foreleg; 11*, midleg; 12*, hindleg. Fig. 13, abdominal segment I of 2nd instar larva in dorsal view. Fig. 14, last abdominal segments of 1st instar larva in dorsal view. Figs. 15, 16, pupa: 15, last segments in lateral view; 16, pupa in dorsal view. Figs. 2, 6, 16 after Schiødte (1864). Figs. 3, 7, 13, 14, 15 after Bertrand (1928b). Fig. 8 after Holmen (1987). Figs. 1, 5, 9 after Vondel (1997).

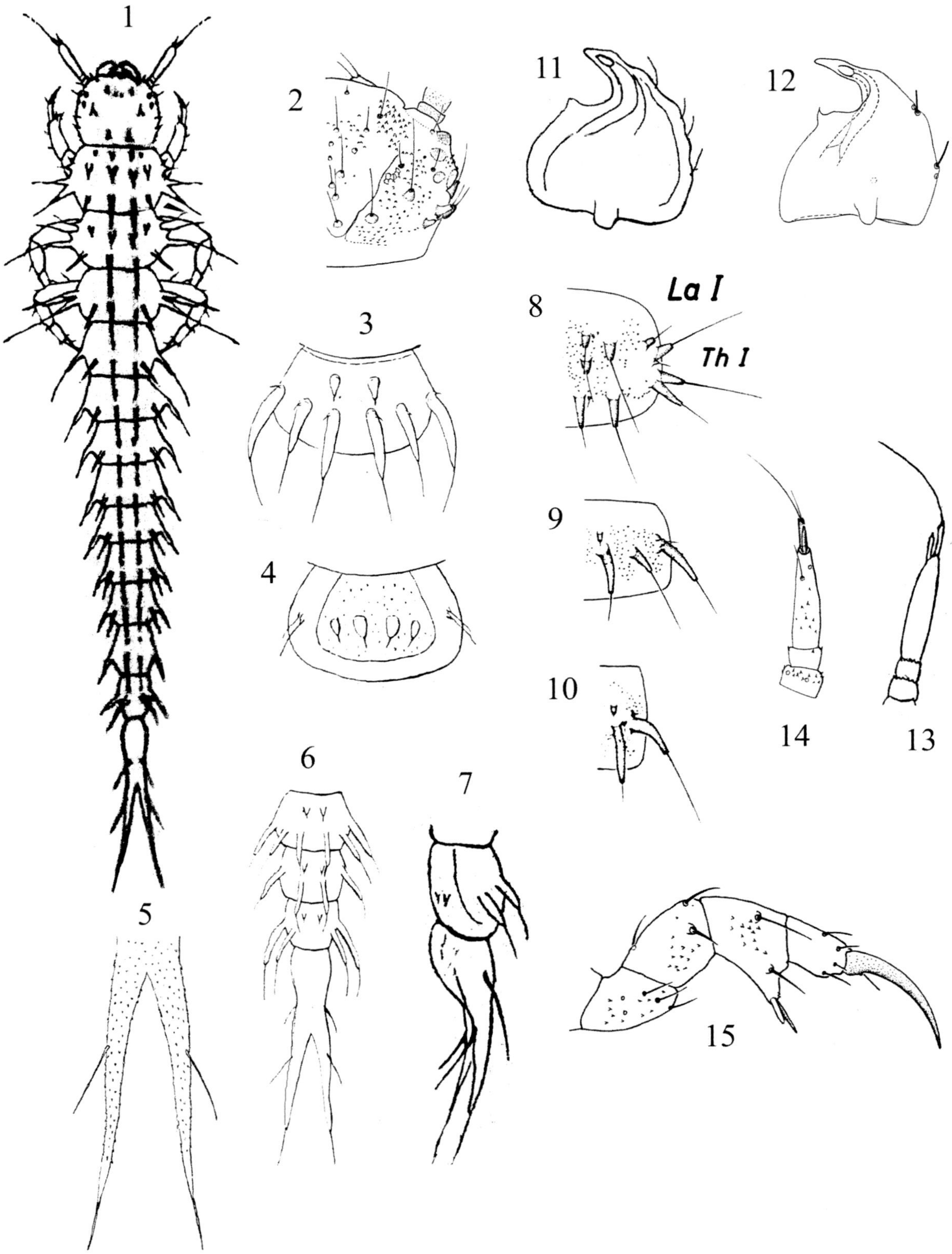

Plate 48. Figs. 1–15. *Haliplus (Neohaliplus) lineatocollis.* 1st instar larva. 1, larva in dorsal view; 2, head in dorsal view (right half); 3 abdominal segment I in dorsal view; 4, abdominal segment I in ventral view; 5, forked part of abdominal segment X; 6 last abdominal segments in dorsal view; 7, last two abdominal segments in lateral view; 8, pronotum; 9, abdominal segment I in dorsal view; 10, abdominal segment IX in dorsal view; 11, 12, mandible; 13, 14, antenna. Figs. 1, 3–7, 11, 13 after Bertrand (1923). Figs. 2, 8–10, 12, 14, 15 after Seeger (1971a).

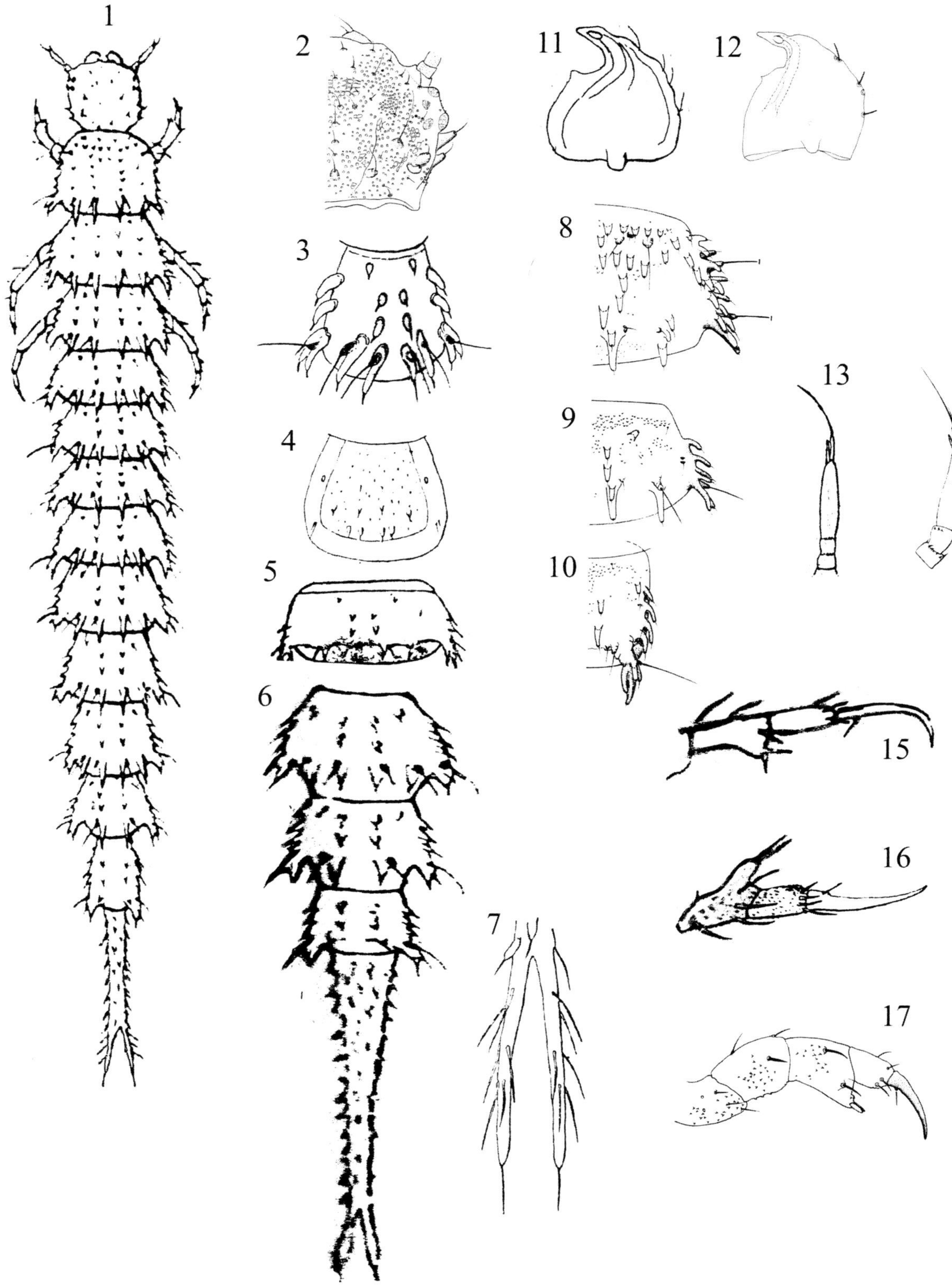

Plate 49. Figs. 1–17. *Haliplus (Neohaliplus) lineatocollis*. 2nd instar larva. 1, larva in dorsal view; 2, head in dorsal view (right half); 3, 5, 9, abdominal segment I in dorsal view; 4, abdominal segment I in ventral view; 6, last abdominal segments in dorsal view; 7, forked part of abdominal segment X; 8, pronotum, 10, abdominal segment X in dorsal view; 11, 12, mandible; 13, 14, antenna; 15–17 last segments of foreleg. Figs. 1, 3, 4, 6, 7, 11, 13, 15, 16 after Bertrand (1923). Fig. 5 after Bertrand (1928b). Figs. 2, 8–10, 12, 14, 17 after Seeger (1971a).

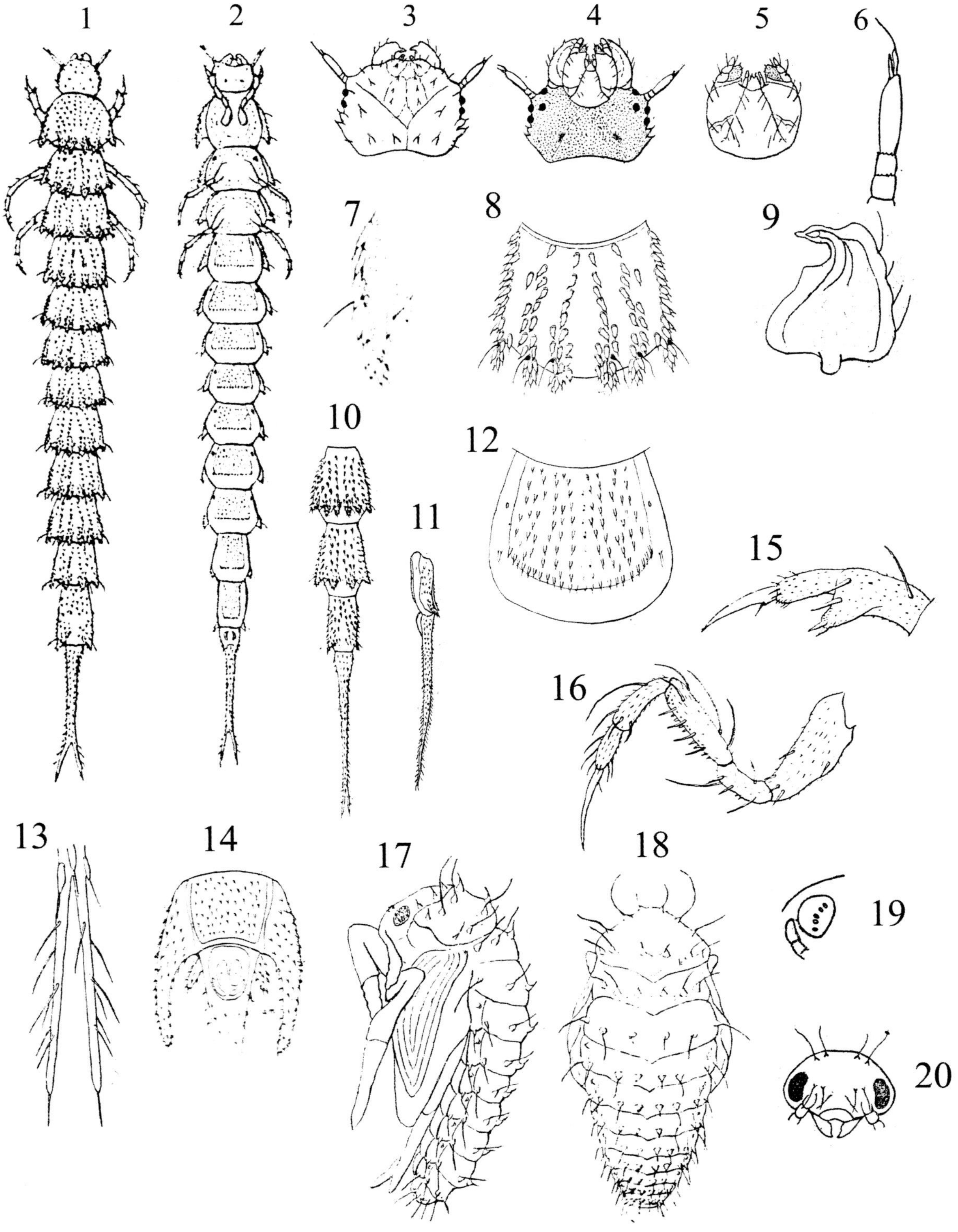

Plate 50. Figs. 1–20. *Haliplus (Neohaliplus) lineatocollis*. 3rd instar larva and pupa. 1, larva in dorsal view; 2, larva in ventral view; 3, head in dorsal view; 4, head in ventral view; 5, labium and maxilla in ventral view, 6, antenna; 7, lateral process of thoracic and abdominal segments; 8, abdominal segment I in dorsal view; 9, mandible; 10, last abdominal segments in dorsal view; 11, last two abdominal segments in lateral view; 12, abdominal segment I in ventral view; 13, forked part of abdominal segment X; 14, abdominal segment IX and base of X in ventral view; 15, last segments of foreleg; 16, hindleg; 17, pupa in lateral view; 18, pupa in dorsal view; 19, ocelli on head of pupa; 20, head of pupa in frontal view. Figs. 7, 14, 19, 20 after Bertrand (1928b). All other Figs. after Bertrand (1923).

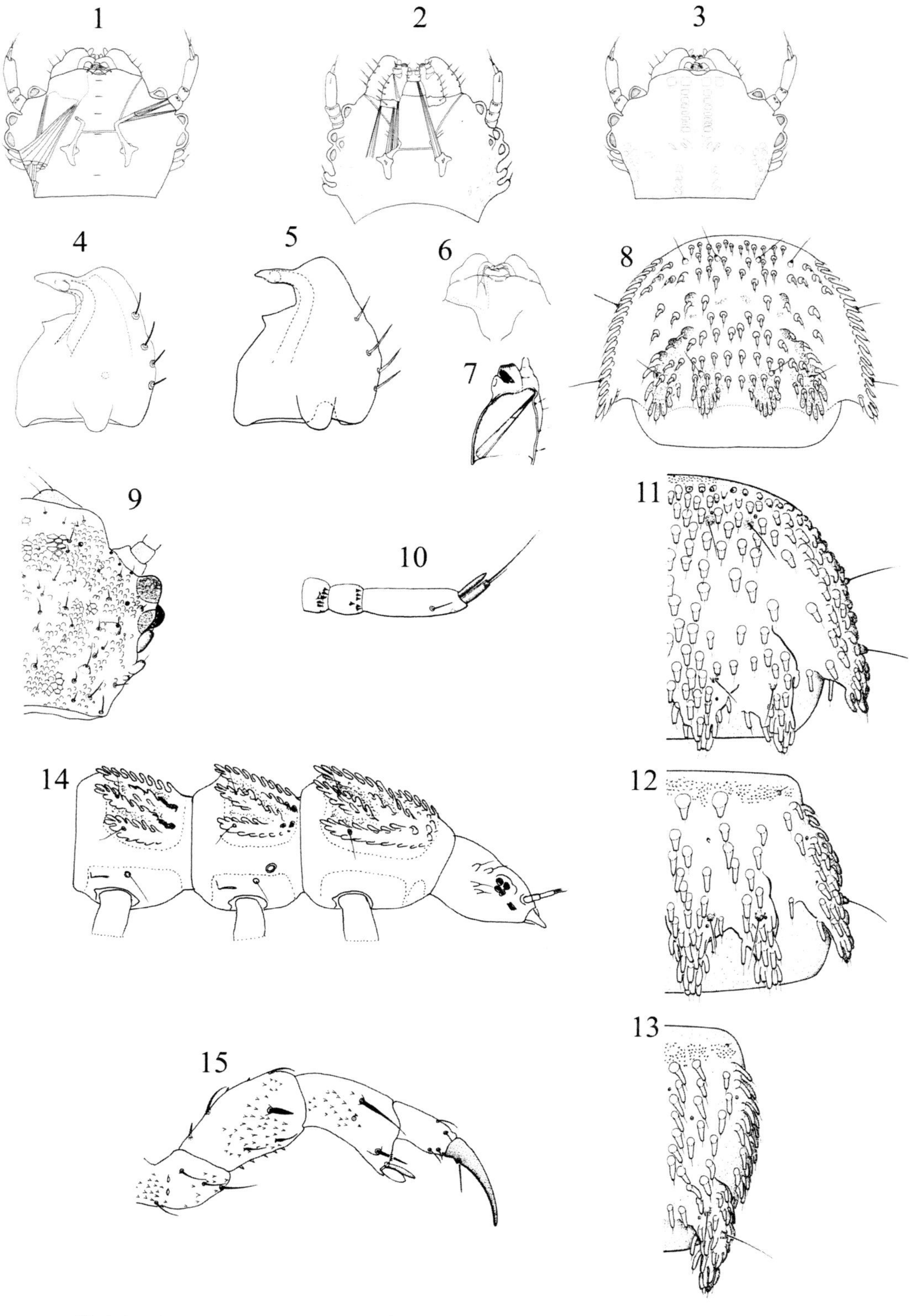

Plate 51. Figs. 1–15. *Haliplus (Neohaliplus) lineatocollis*. 3rd instar larva (continued). 1, head, horizontal cross section showing mandible– and antenna muscles; 2, head, horizontal cross section showing maxilla– and labium muscles; 3, head in dorsal view; 4, 5, mandible; 6, mandibles in dorsal view; 7, maxilla in dorsal view; 8, 11, pronotum; 9, head in dorsal view (right half); 10, antenna; 12, abdominal segment I in dorsal view (right half); 13, abdominal segment IX in dorsal view (right half); 14, head and thorax in lateral view; 15, foreleg. Figs. 4, 9, 10–13, 15 after Seeger (1971a). Figs. 1–3, 5–7 after Beutel (1986). Figs. 8, 14 after Vondel (1997).

Plate 52. Figs. 1–8. *Haliplus* sp. Figs. 1, 2, *Haliplus (Nipponiplus) japonicus*, 3rd instar larva in dorsal view. Figs. 3–8, *Haliplus (Haliplus) kamiyai*, 3–5, 3rd instar larva in dorsal, lateral and ventral view; 6, larva feeding on Spyrogira; 7, foreleg of 3rd instar larva; 8, pupa. Fig. 1 after Mitamura (2017). Fig. 2 after Nakajima (2020). Figs. 3–8 after Watanabe & Yamasaki (2020).

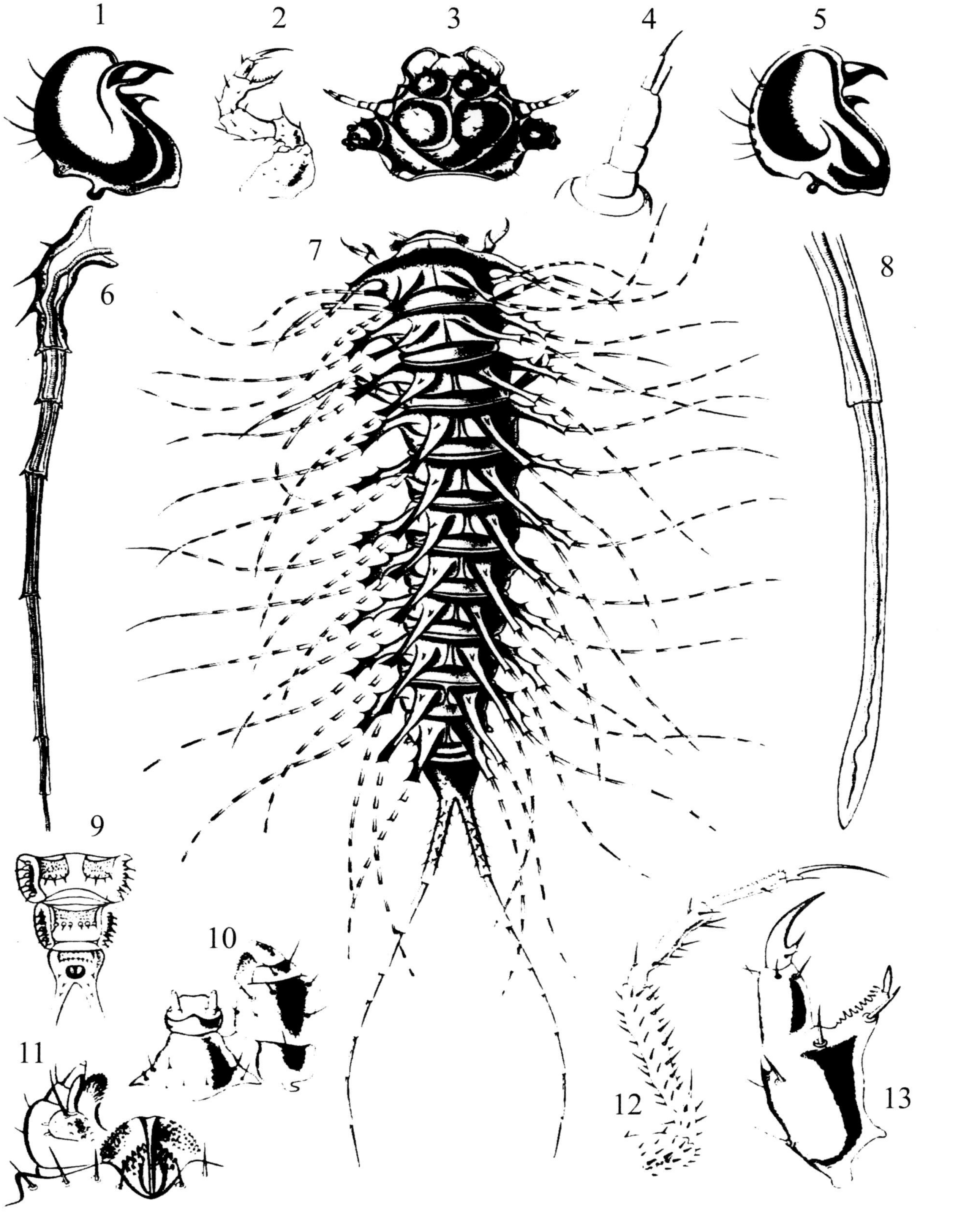

Plate 53. Figs. 1–13. *Peltodytes (Peltodytes) caesus*. 3rd instar larva. 1, left mandible in dorsal view; 2, foreleg; 3, head in dorsal view; 4, antenna, 5, right mandible in ventral view; 6, filamentous dorsal and lateral process; 7 larva in dorsal view; 8, end of filamentous dorsal and lateral process; 9, abdominal segment VII, VIII and IX in ventral view; 10, maxilla and labium in dorsal view; 11, maxilla and labium in ventral view; 12, hindleg; 13, claw, tarsus and tibia of foreleg. All Figs. after Schiødte (1872).

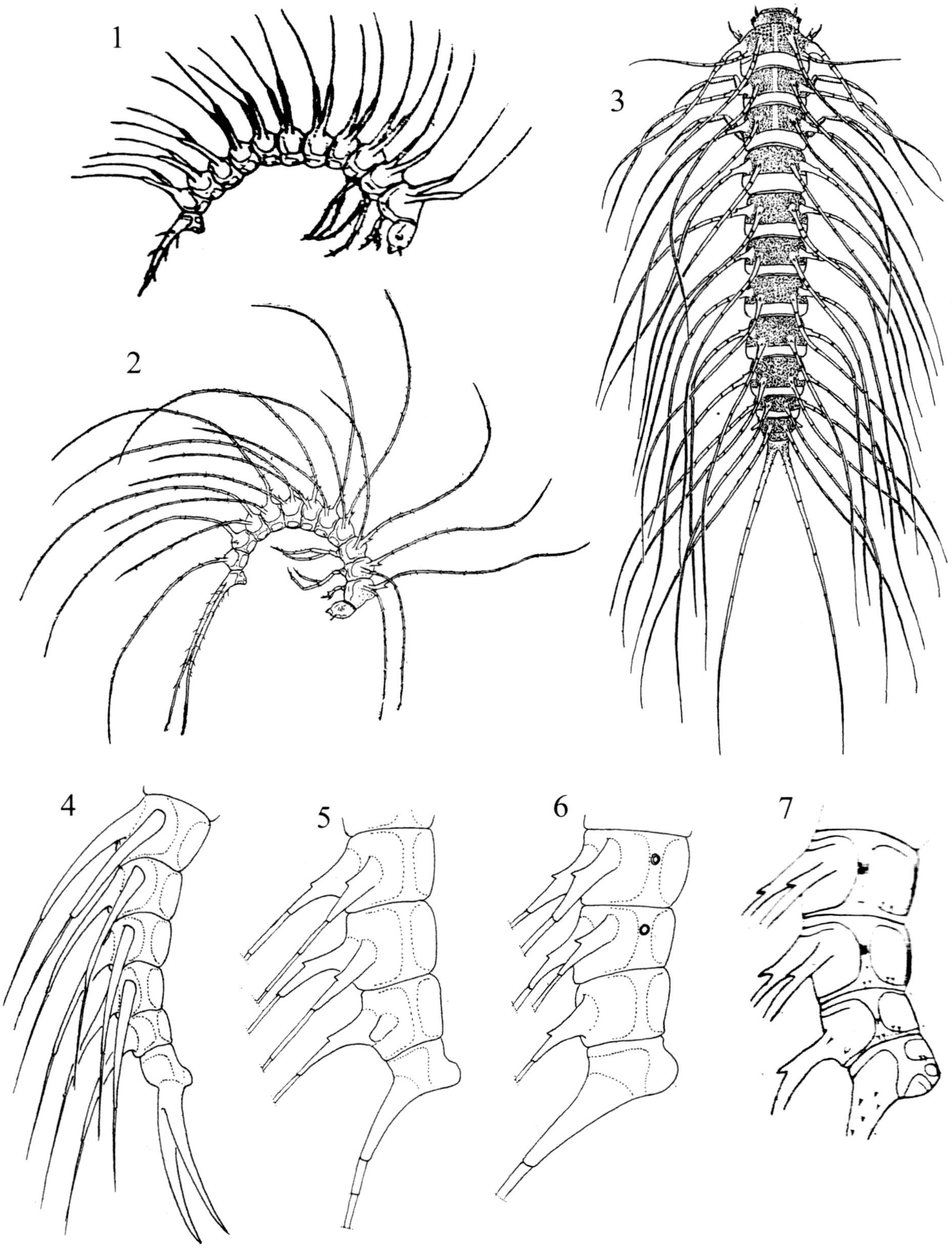

Plate 54. Figs. 1–7. *Peltodytes (Peltodytes) caesus*. 1, 1st instar larva in lateral view; 2, 2nd instar larva in lateral view; 3, 3rd instar larva in dorsal view; 4, last abdominal segments of 1st instar larva in lateral view; 5, last abdominal segments of 2nd instar larva in lateral view; 6, 7, last abdominal segments of 3rd instar larva in lateral view. Figs. 1, 2, 7 after Bertrand (1928b). Figs. 3, 4–6 after Vondel (1997).

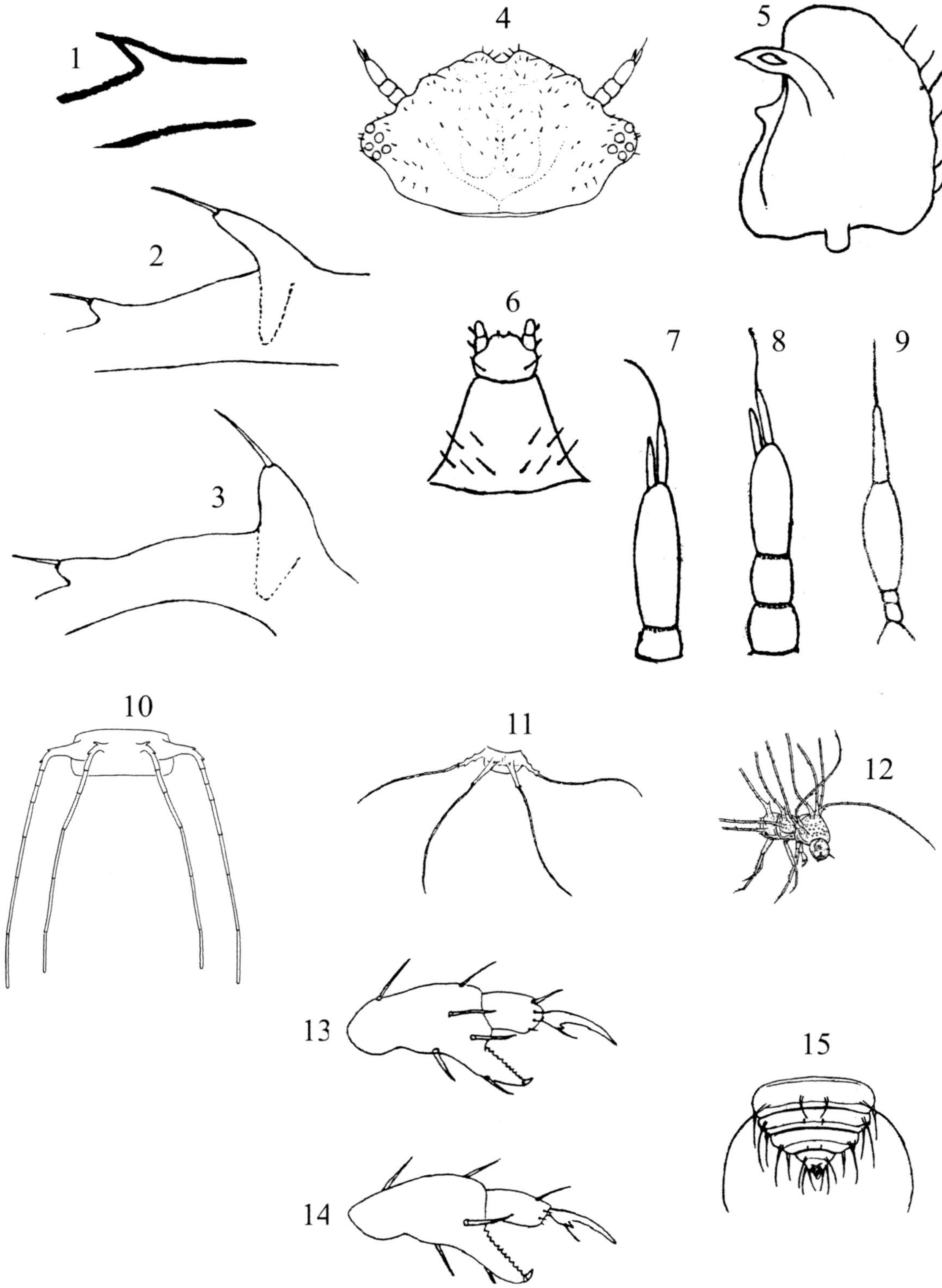

Plate 55. Figs. 1, 2, 4–6, 8–14, *Peltodytes (Peltodytes) caesus*, 3rd instar larva. 1, 2, basal tooth on dorsal and lateral filamentous process; 4, head in dorsal view; 5, mandible; 6, labium; 8, 9 antenna; 10, 11, abdominal segment I in dorsal view; 12, larva in frontal view; 13, claw, tarsus and tibia of foreleg in dorsal view; 14, claw, tarsus and tibia in ventral view. Fig. 7. *Peltodytes (Peltodytes) caesus*, antenna of 2nd instar larva. Fig. 15. *Peltodytes (Peltodytes) caesus*, last segments of pupa in dorsal view. Fig. 3. *Peltodytes (Peltodytes) rotundatus*, basal tooth on dorsal and lateral filamentous process. Fig. 11 after Bertrand (1923). Figs. 1, 4–8, 11–14, 15 after Bertrand (1928b). Figs. 2, 3 after Bertrand (1951). Fig. 9 after Steinmann & Zombori (1984). Fig. 10 after Holmen (1987).

Plate 56. Figs. 1–8. *Peltodytes (Peltodytes) intermedius*. Eggs, 3rd instar larva and pupa. 1, 2, 4, 7 3rd instar larva in dorsal view; 3, 3rd instar larva in lateral view; 5, eggs; 6, 8, pupa. Fig. 9. *Peltodytes (Peltodytes) sinensis*, 3rd instar larva. Fig. 1 after Sato & Yoshitomi (2018). Fig. 2 after Fukuda et al. (1959). Fig. 4 after Mitamura (2017). Figs. 5, 6, 7 photographed by Kitano (pers. comm.). Figs. 3, 8 photographed by Hayashi (pers. comm.). Fig. 9 after Jia et all. (2024) (photo by Zuqi Mai).

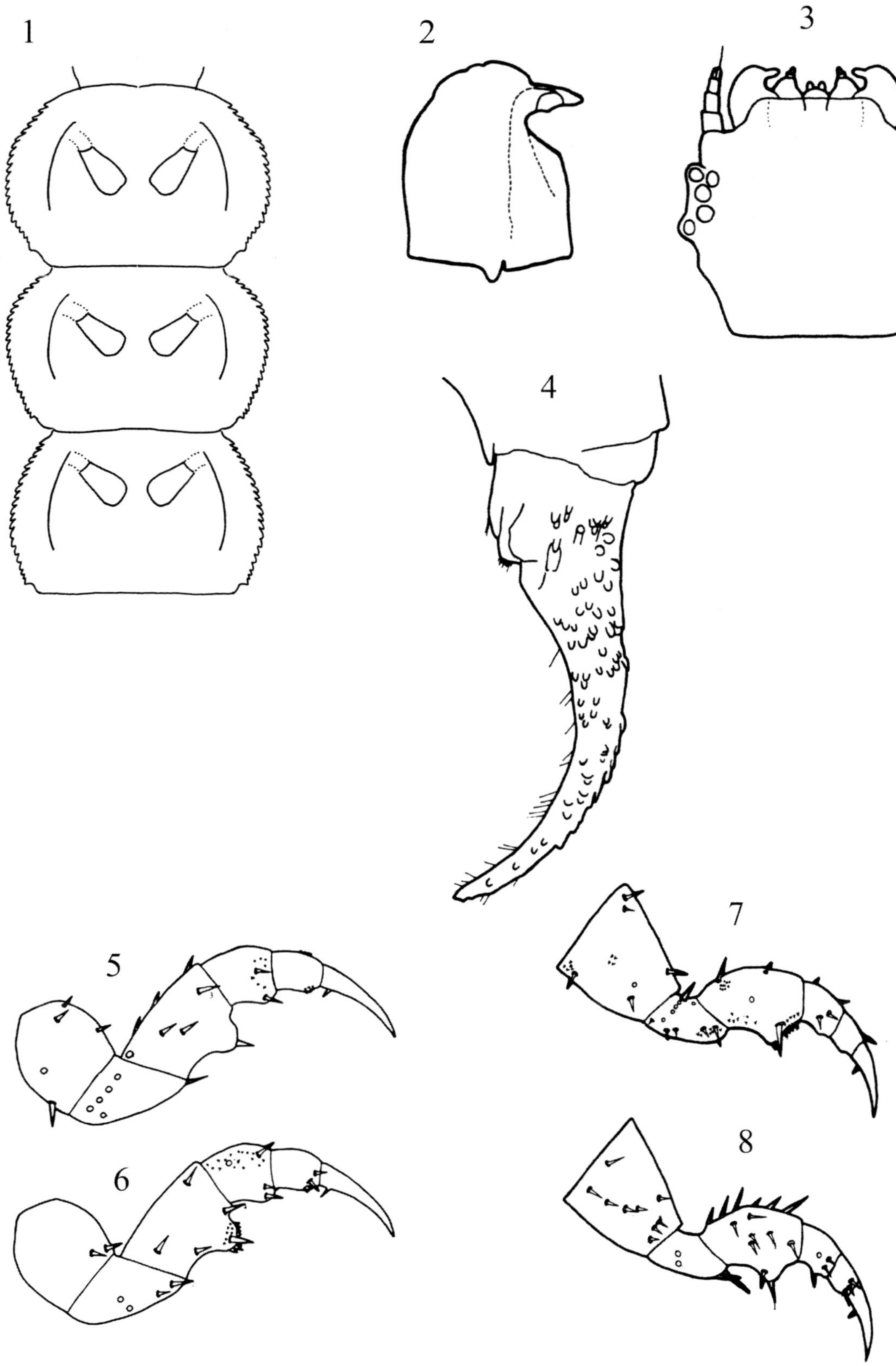

Plate 57. Figs. 1–8. *Brychius hornii*. Figs. 5, 6, foreleg of 2nd instar larva, in dorsal and ventral view. Figs. 1–4, 7, 8, 3rd instar larva: 1 thorax in ventral view; 2, mandible; 3, head in dorsal view; 4, abdominal segment X in lateral view; 7, 8, foreleg in dorsal and ventral view. All Figs. after Mousseau (2004). Fig. 1 redrawn after Mousseau (2004).

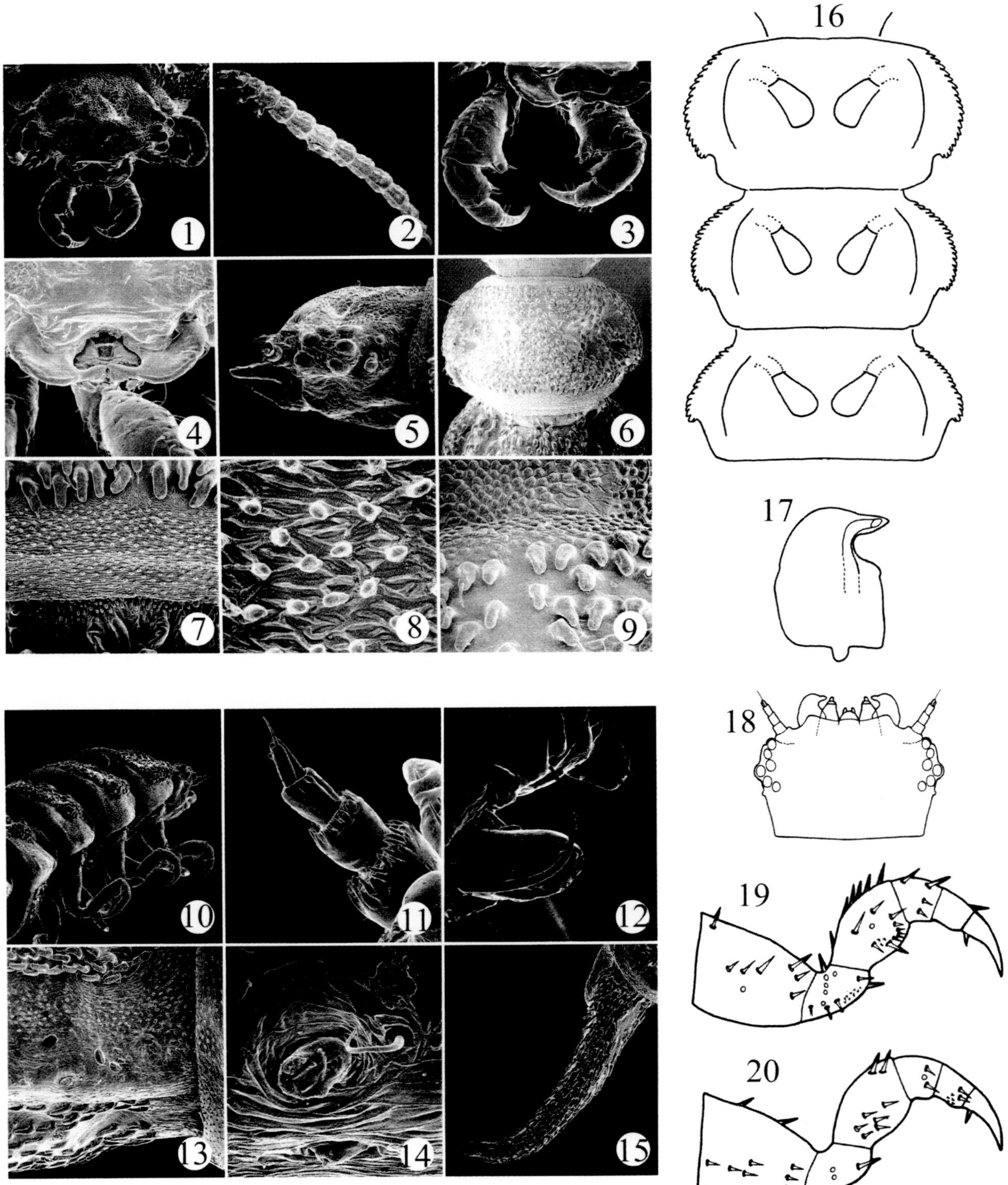

Plate 58. Figs. 1–20. *Brychius hungerfordi*. 3rd instar larva. 1, head in dorsal view; 2, larva in lateral view; 3, forelegs and mandibles in dorsal view; 4, mandibles in ventral view; 5, head in lateral view; 6, pronotum dorsal sculpture; 7, pronotal base, asperities and granular intersegmental sculpture; 8, granular intersegmental sculpture; 9, pronotal apex, granular intersegmental sculpture and asperities; 10, head, thorax and legs in oblique view; 11, antenna in dorsal view; 12, labial palpus, apical segment (above), and maxillar palpus, apical segment (below); 13, spiracle, abdominal segment II, lateral view; 14, spiracle; 15, abdominal segment X in lateral view; 16, thorax in ventral view; 17, mandible; 18, head in dorsal view; 19, foreleg in dorsal view; 20, foreleg in ventral view. Figs. 1–15 after Strand & Spangler (1994). Figs. 16–20 after Mousseau (2004). Fig. 16 redrawn after Mousseau (2004).

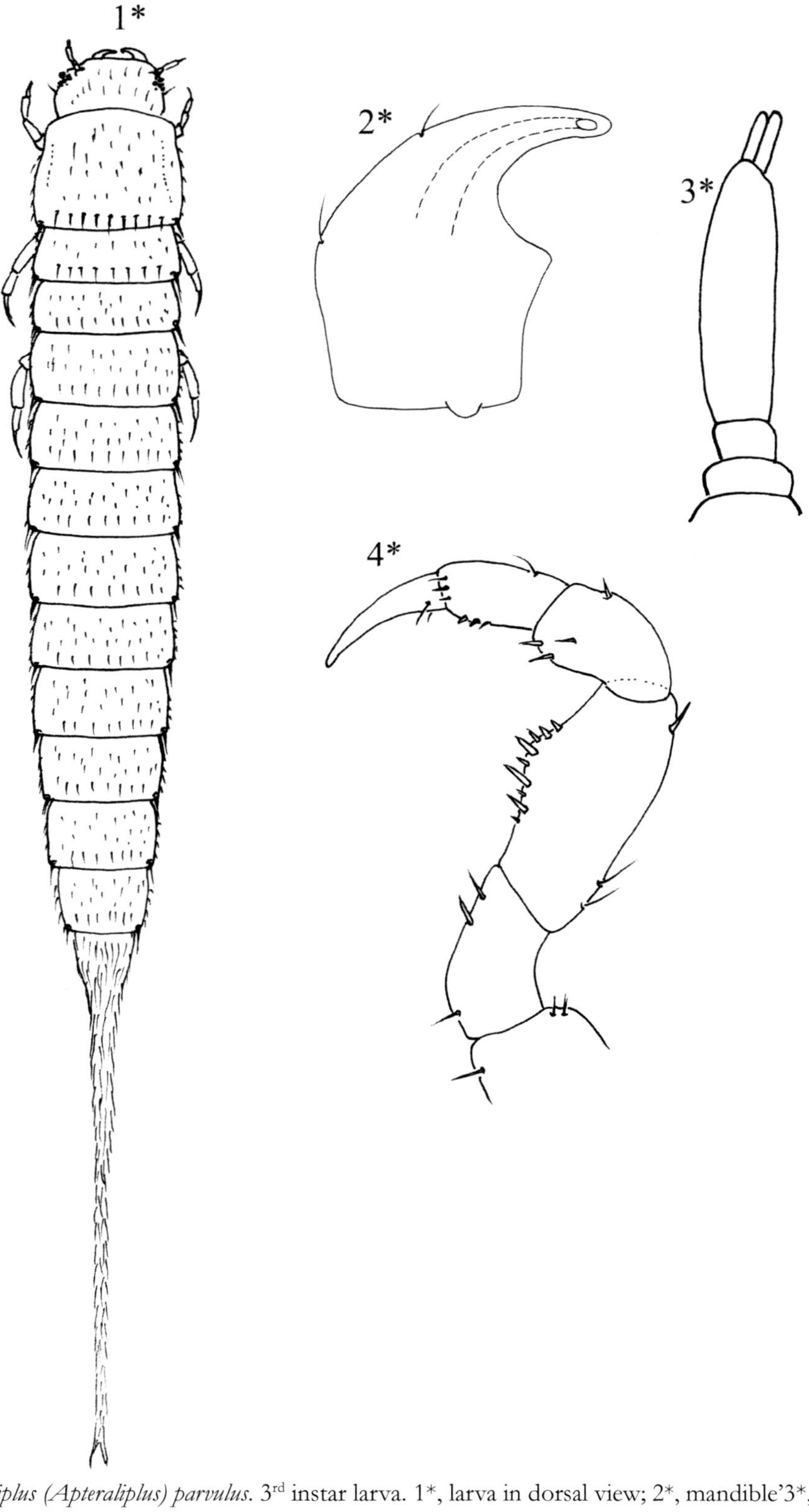

Plate 59. Figs. 1*–4*. *Haliplus (Apteraliplus) parvulus*. 3rd instar larva. 1*, larva in dorsal view; 2*, mandible'3*, antenna; 4*, foreleg.

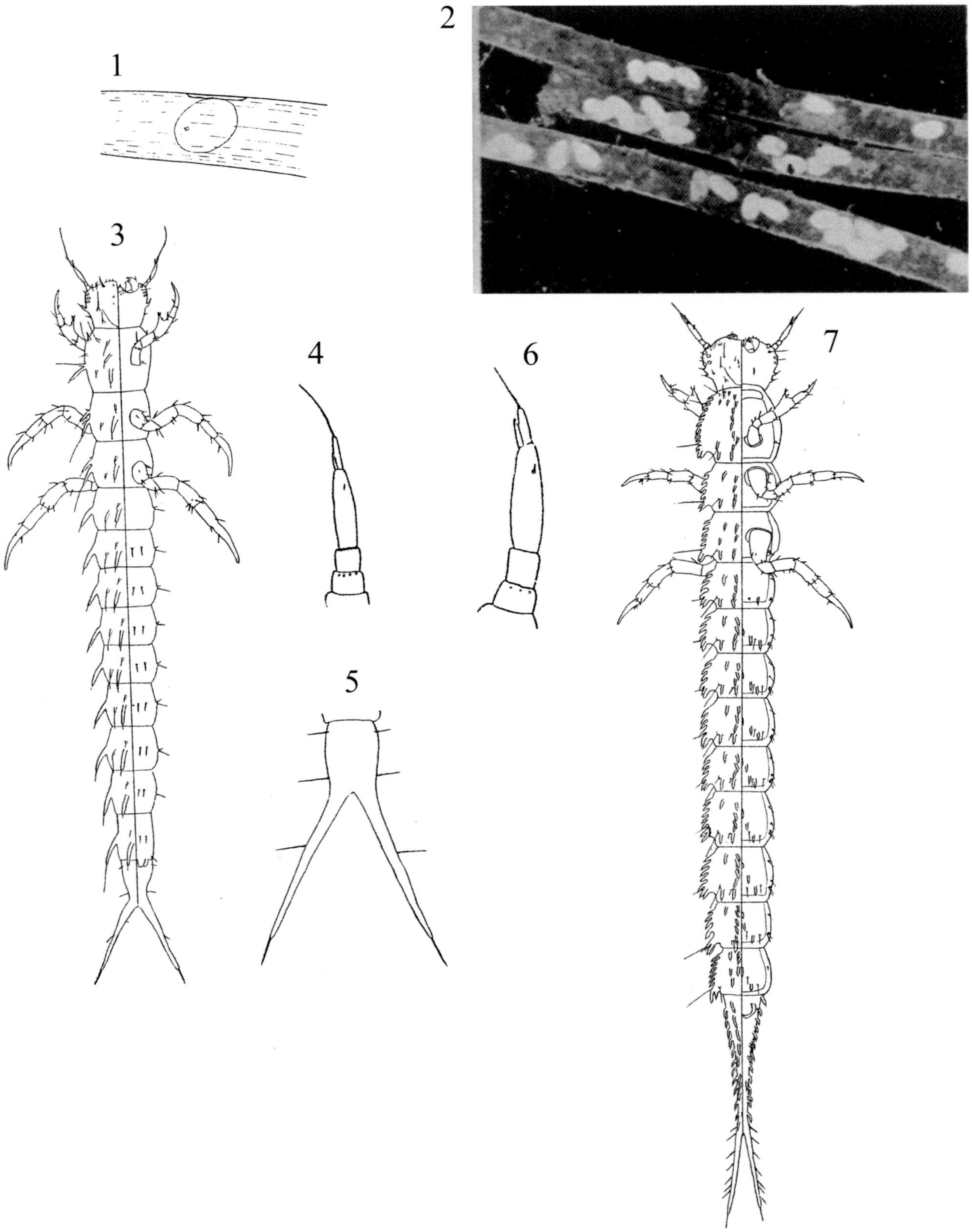

Plate 60. Figs. 1–7. *Haliplus (Haliplus) immaculicollis.* 1, egg; 2, eggs in algae; 3, 1st instar larva, left in dorsal and right ventral view; 4, antenna of 1st instar; 5, abdominal segment X of 1st instar; 6, antenna of 2nd instar; 7, 2nd instar larva, left in dorsal view, right in ventral view. Fig. 2 after Matheson (1912). Other Figs. after Hickman (1930).

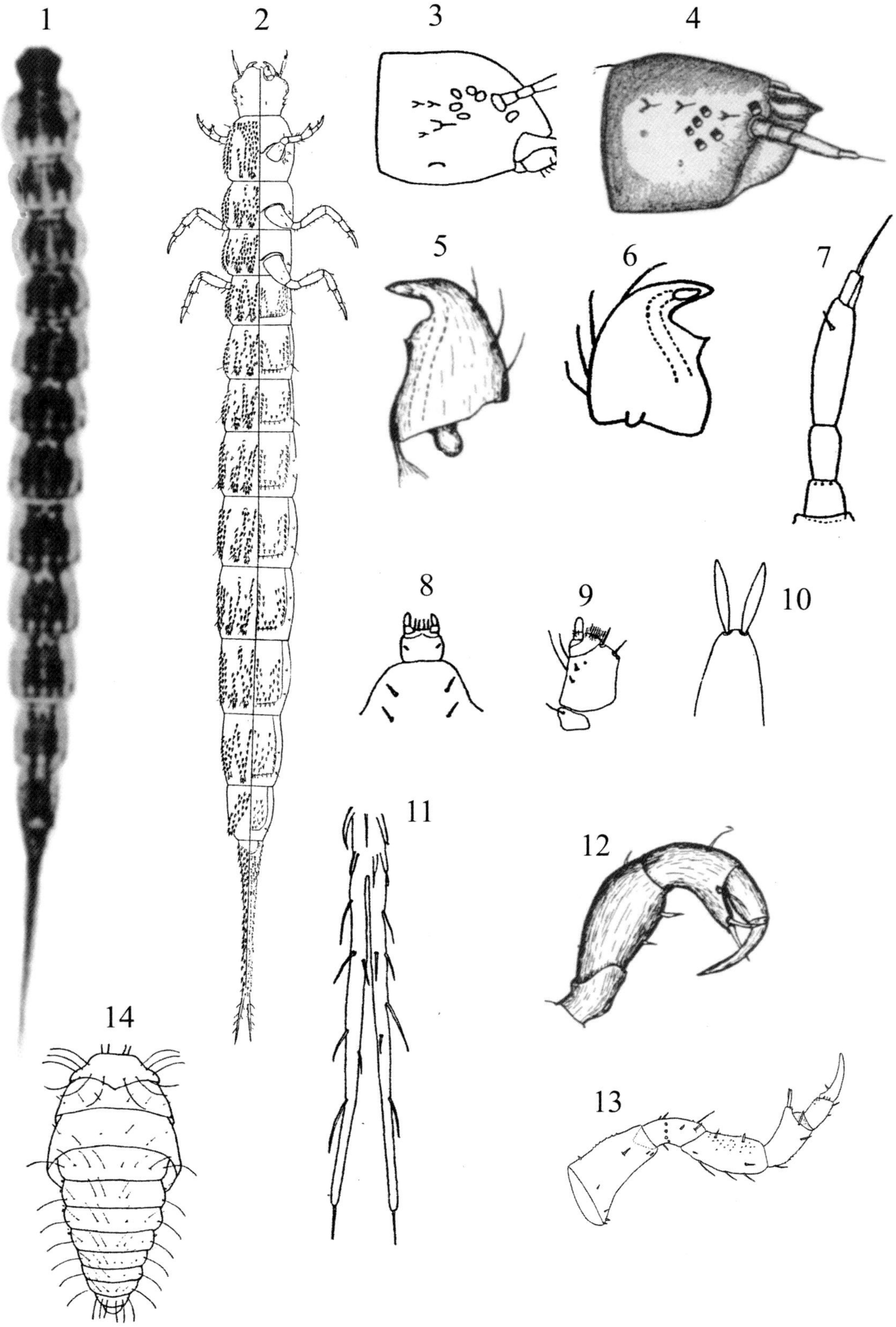

Plate 61. Figs. 1–14. *Haliplus (Haliplus) immaculicollis*. Figs. 1–13. 3rd instar larva. 1, body in dorsal view; 2, larva, left in dorsal view, right in ventral view; 3, 4, head in lateral view; 5, 6, mandible; 7, antenna; 8, labium; 9, maxilla; 10, prolongation of tibia of foreleg; 11, forked end of abdominal segment X; 12, 13, foreleg. Fig. 14, pupa in dorsal view. Figs. 1, 4, 5, 12 after Matheson (1912). Figs. 2, 3, 6–11, 13, 14 after Hickman (1930).

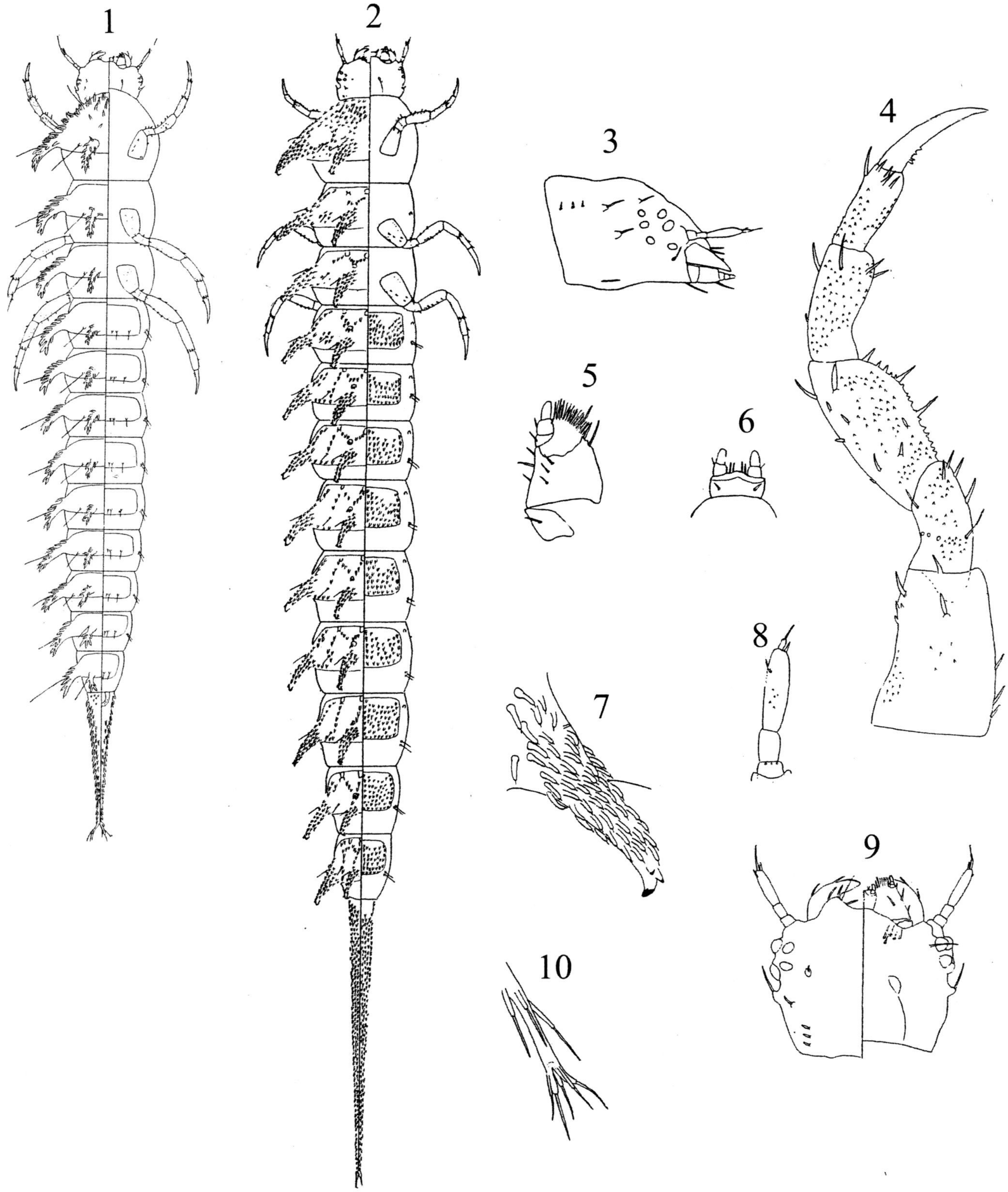

Plate 62. Fig. 1–10. *Haliplus (Liaphlus) cribrarius.* Fig. 1, 2nd instar larva, left in dorsal view, right in ventral view. Figs. 2–10, 3rd instar larva, 2, larva, left in dorsal view, right in ventral view; 3, head in lateral view; 4, foreleg; 5, maxilla; 6, labium; 7, lateral process of thoracic and abdominal segments; 8, antenna; 9, head, left in dorsal view, right in ventral view; 10, end of abdominal segment X. All Figs. after Hickman (1930).

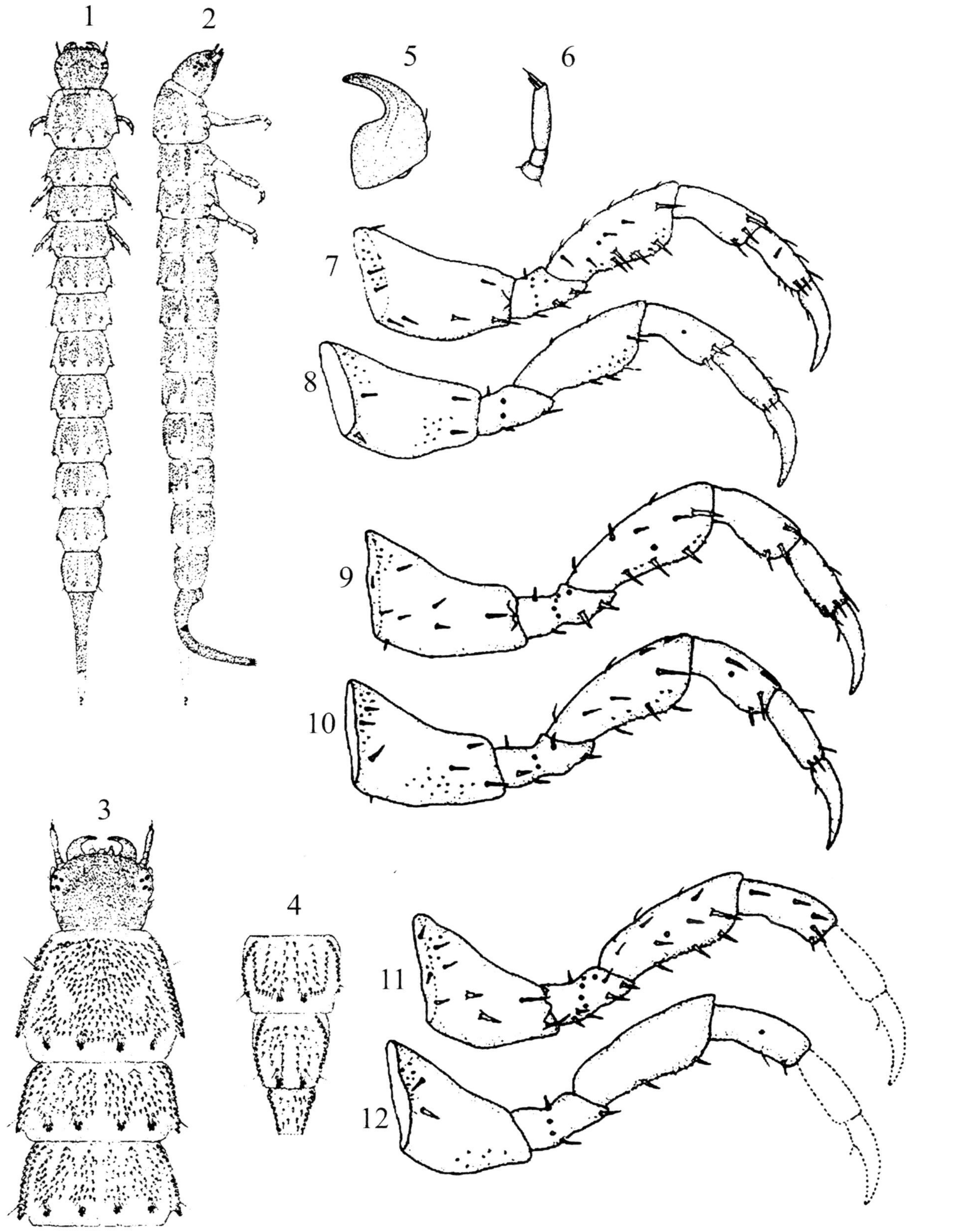

Plate 63. Figs. 1–12. *Haliplus (Paraliaphlus) deceptus*. 3rd instar larva. 1, larva in dorsal view; 2, larva in lateral view; 3, head and thorax in dorsal view; 4, abdominal segment VIII and IX in dorsal view; 5, mandible; 6, antenna; 7, foreleg in dorsal view; 8, foreleg in ventral view; 9, midleg in dorsal view; 10, midleg in ventral view; 11, hindleg in dorsal view; 12, hindleg in ventral view. All Figs. after Vondel (2011).

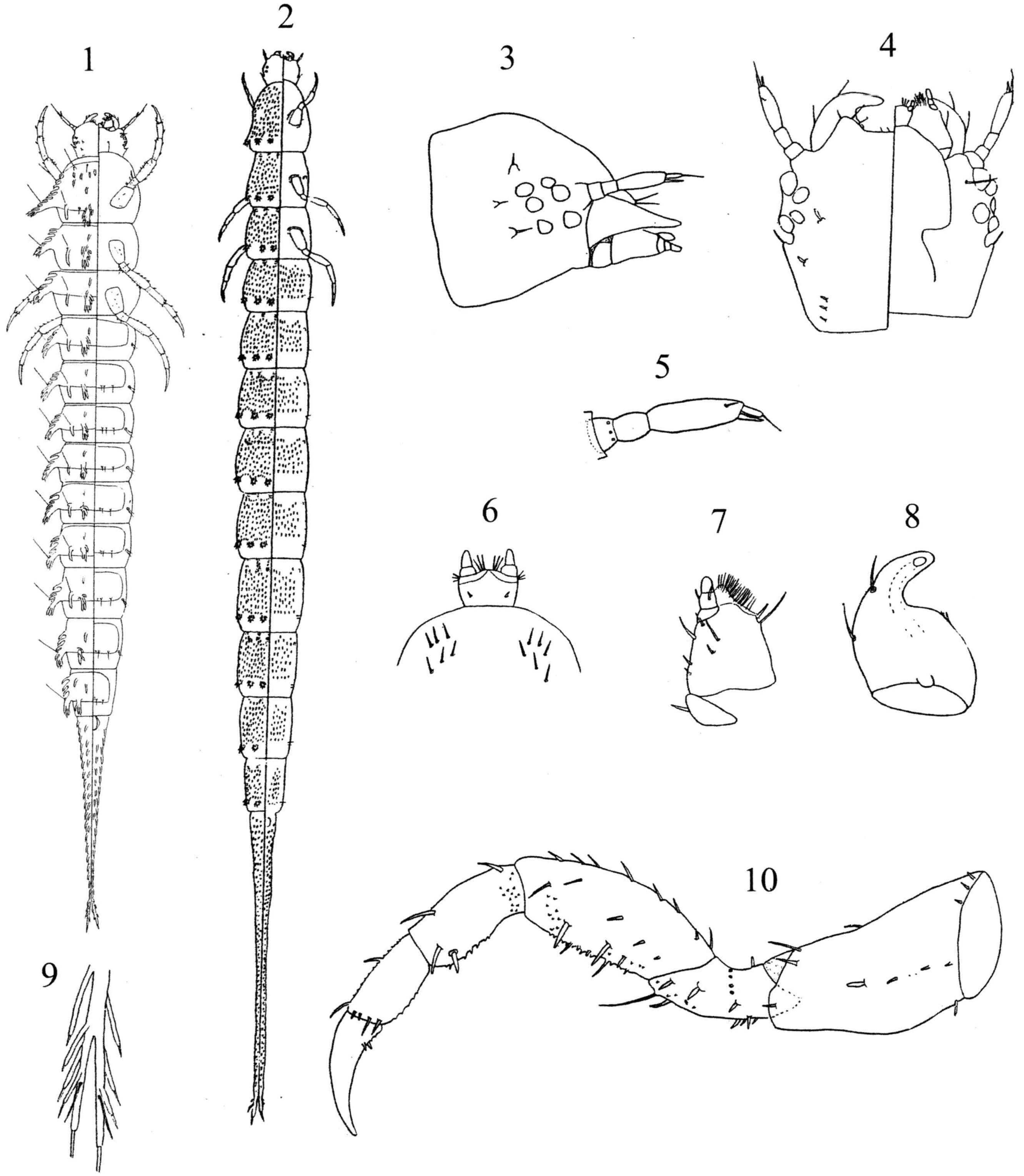

Plate 64. Figs. 1–10. *Haliplus (Paraliaphlus) triopsis*. Fig. 1, 2nd instar larva, left in dorsal view, right in ventral view. Figs. 2–10. 3rd instar larva, 2, larva, left in dorsal view, right in ventral view; 3, head in lateral view; 4, head, left in dorsal view, right in ventral view; 5, antenna; 6, labium; 7, maxilla; 8, mandible; 9, forked end of abdominal segment X; 10, foreleg. All Figs. after Hickman (1930).

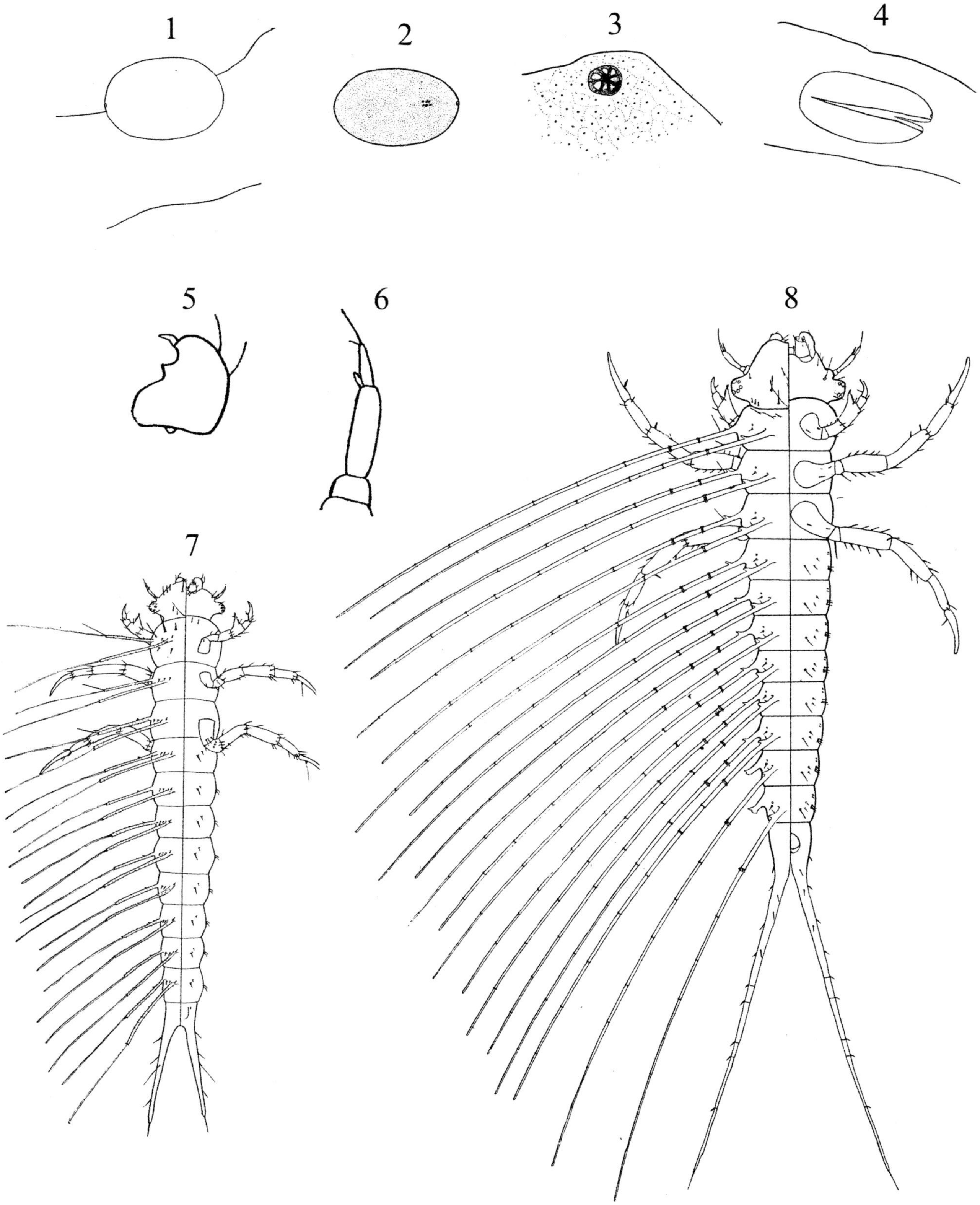

Plate 65. Figs. 1–8. *Peltodytes (Neopeltodytes) duodecimpunctatus.* Figs. 1–4, egg. 1, egg attached to branch of Ceratophyllum; 2, egg showing group of ocelli; 3, micropyle; 4, egg after hatching. Figs. 5–7, 1st instar larva. 5, mandible; 6, antenna; 7 larva, left in dorsal view, right in ventral view. Fig. 8, 2nd instar larva, left in dorsal view, right in ventral view. All Figs. after Hickman (1930, as *P. lengi*, jun. syn. of *P. duodecimpunctatus*).

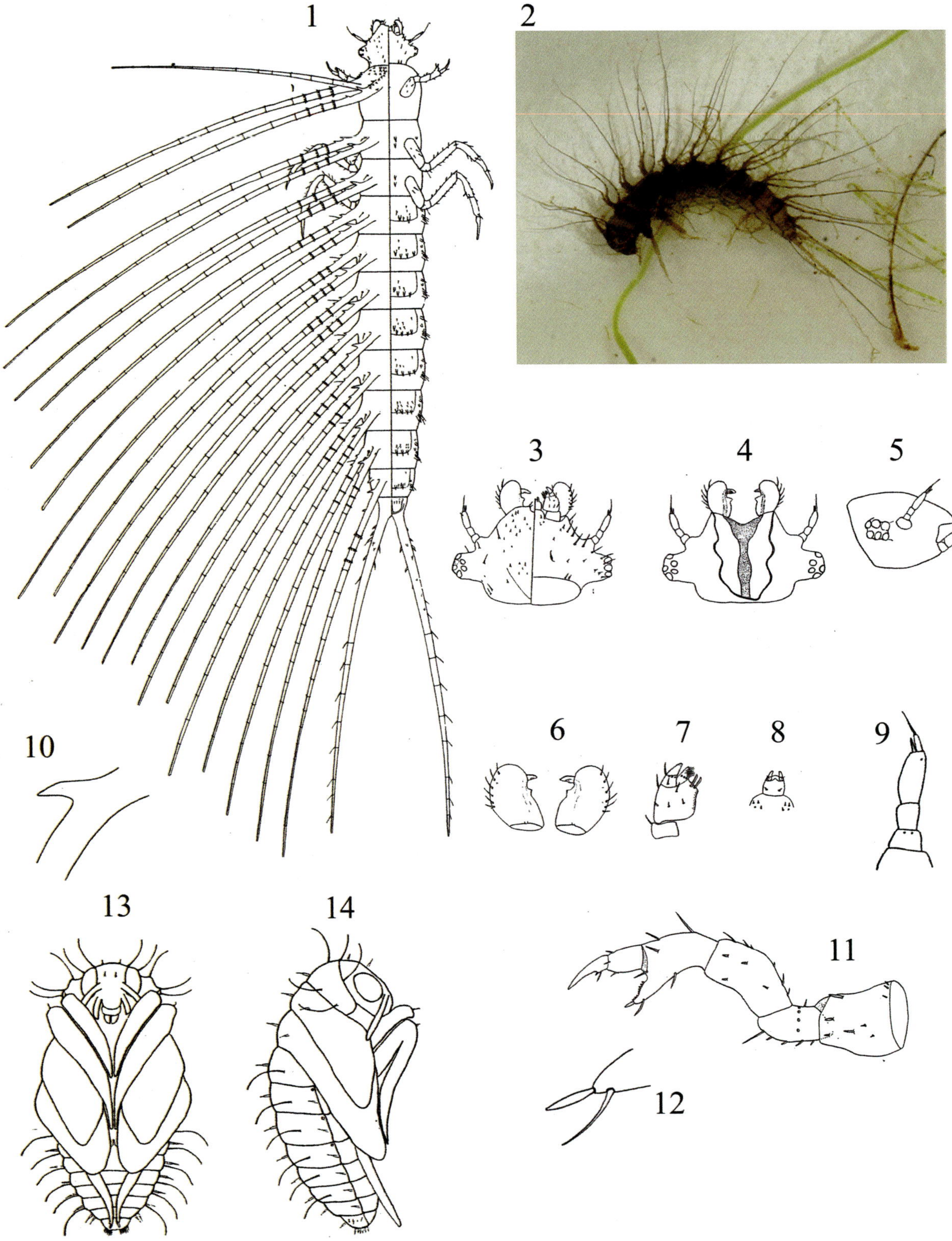

Plate 66. Figs. 1–14. *Peltodytes (Neopeltodytes) duodecimpunctatus*. Figs. 1–12, 3rd instar larva, 1, larva, left in dorsal view, right in ventral view; 2, larva in lateral view; 3, head, left in dorsal view, right in ventral view; 4, head capsule showing suction canal of mandible; 5, head in lateral view; 6, mandibles in dorsal view; 7, maxilla; 8, labium; 9, antenna; 10, basal tooth on filamentous dorsal and lateral process; 11, foreleg; 12, top of tibial extension. Figs. 13, 14, pupa, 13, in ventral view; 14, in lateral view. Fig. 2 after Chandler (2008). Other Figs. after Hickman (1930).

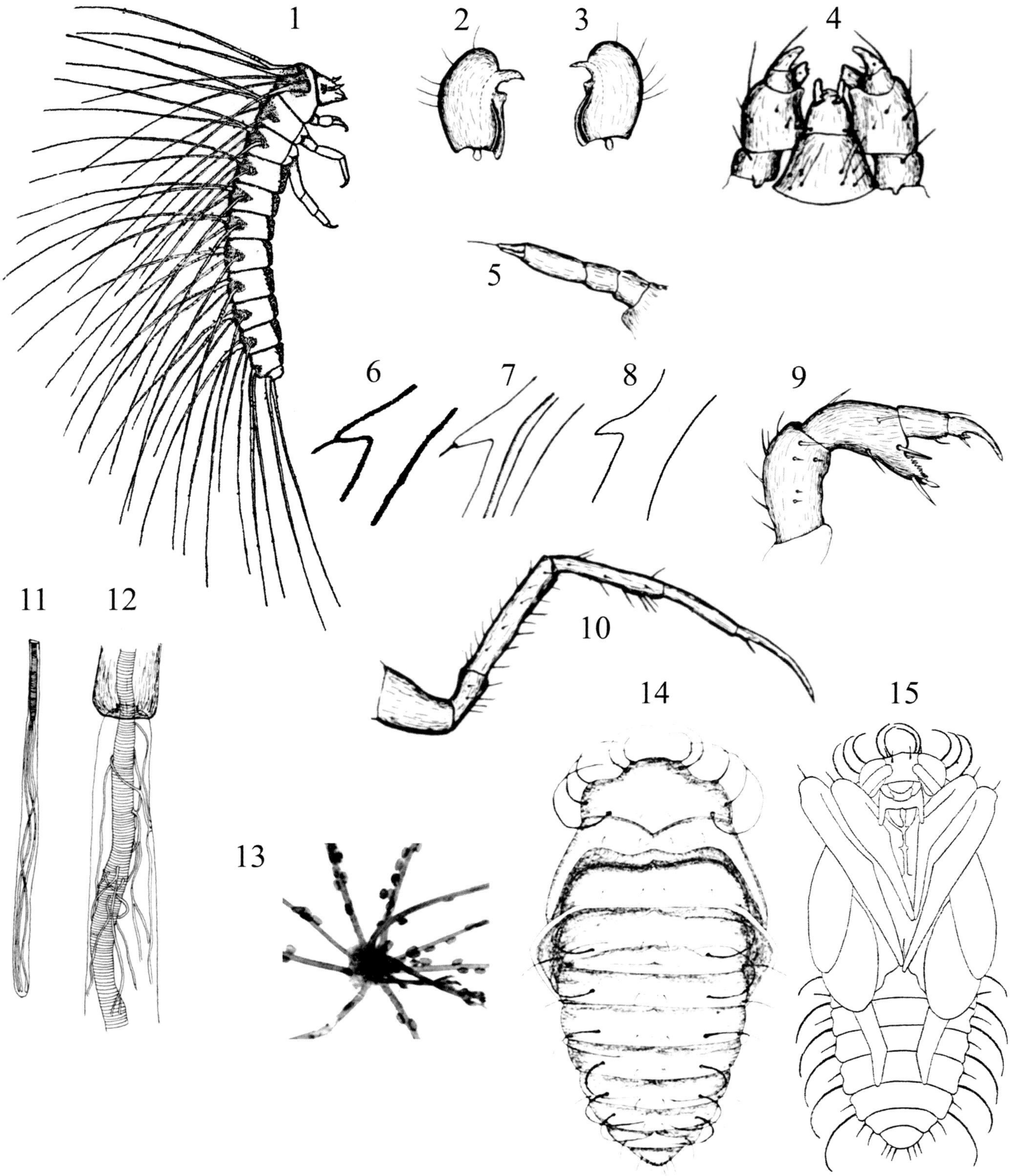

Plate 67. Figs. 1–15. *Peltodytes (Neopeltodytes) edentulus*. Fig. 13, eggs on Chara . Figs. 1–12. 3rd instar larva, 1, larva in lateral view; 2, left mandible; 3, right mandible; 4, labium and maxillae; 5, antenna; 6–8, basal tooth on lateral and dorsal filamentous processes (or spine); 9, foreleg; 10, hindleg; 11, termination of a trachea in three tracheoles in the terminal joint of a larval spine; 12, portion of a dorsal spine showing the trachea and its tracheoles. Figs. 14–15, pupa in dorsal and ventral view. Figs. 2–5, 7, 9–13, 14 after Matheson (1912). Figs. 1, 15 after Wilson (1924). Fig. 6 after Bertrand (1928b). Fig. 8 after Hickman (1930).

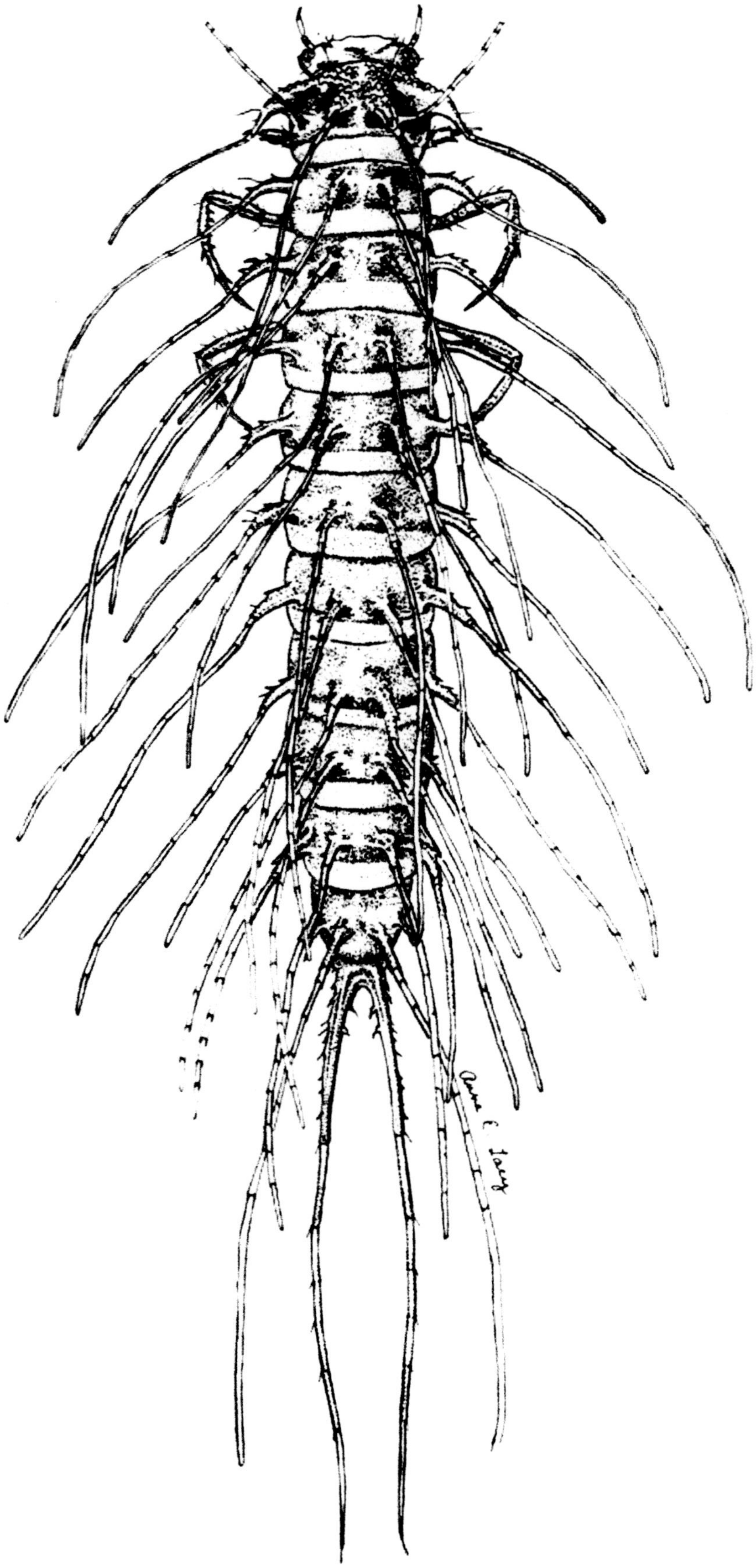

Plate 68. *Peltodytes (Neopeltodytes) litoralis.* 3rd instar larva in dorsal view. After Orrell (2012, drawing by Anna C. Lacey).

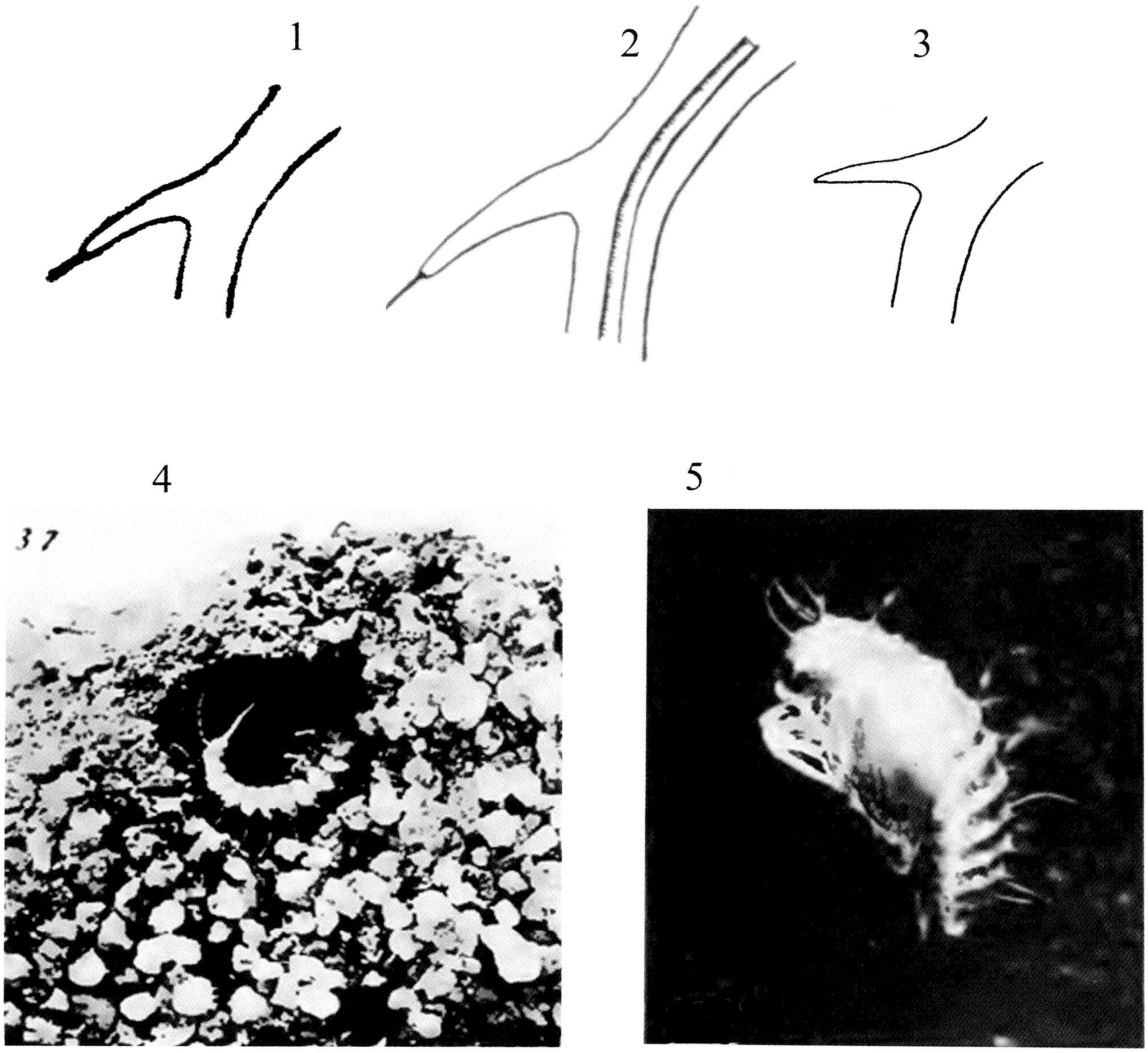

Plate 69. Figs. 1, 2, 4, 5, *Peltodytes (Neopeltodytes) muticus*, 3rd instar larva. 1, 2, basal tooth on filamentous dorsal and lateral process; 4, larva in pupation chamber; 5, pupa. Fig. 3, *Peltodytes (Neopeltodytes) sexmaculatus*, basal tooth on filamentous dorsal and lateral process of 3rd instar larva. Figs. 2, 4, 5 after Matheson (1912). Fig. 1 after Bertrand (1928b). Fig. 3 after Hickman (1930).

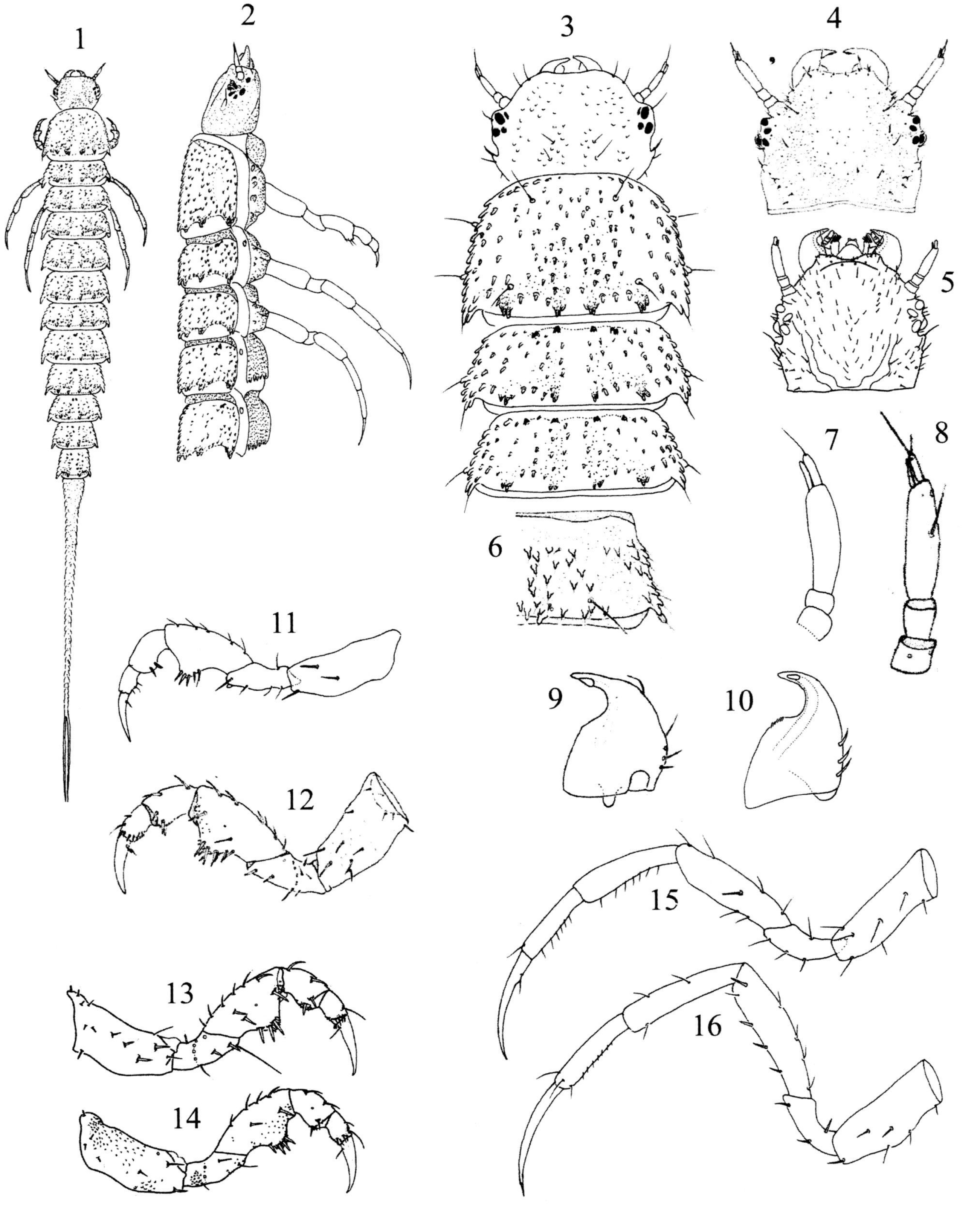

Plate 70. Figs. 1–16. *Haliplus (Haliplidius) subseriatus*. 3rd instar larva. 1, larva in dorsal view; 2, head, thorax and two first abdominal segments in lateral view; 3, head and thorax in dorsal view; 4, 5, head in dorsal view; 6, abdominal segment I in dorsal view (right half); 7, 8, antenna; 9, 10, mandible; 11, 12, 13, foreleg in dorsal view; 14, foreleg in ventral view; 15, midleg; 16, hindleg. Figs. 1–3, 7, 10, 11, 15, 16 after Vondel (2001). Figs. 4, 6, 8, 9, 12 after Archangelsy & Michat (2014). Figs. 5, 13, 14 after Michat et al. (2020).

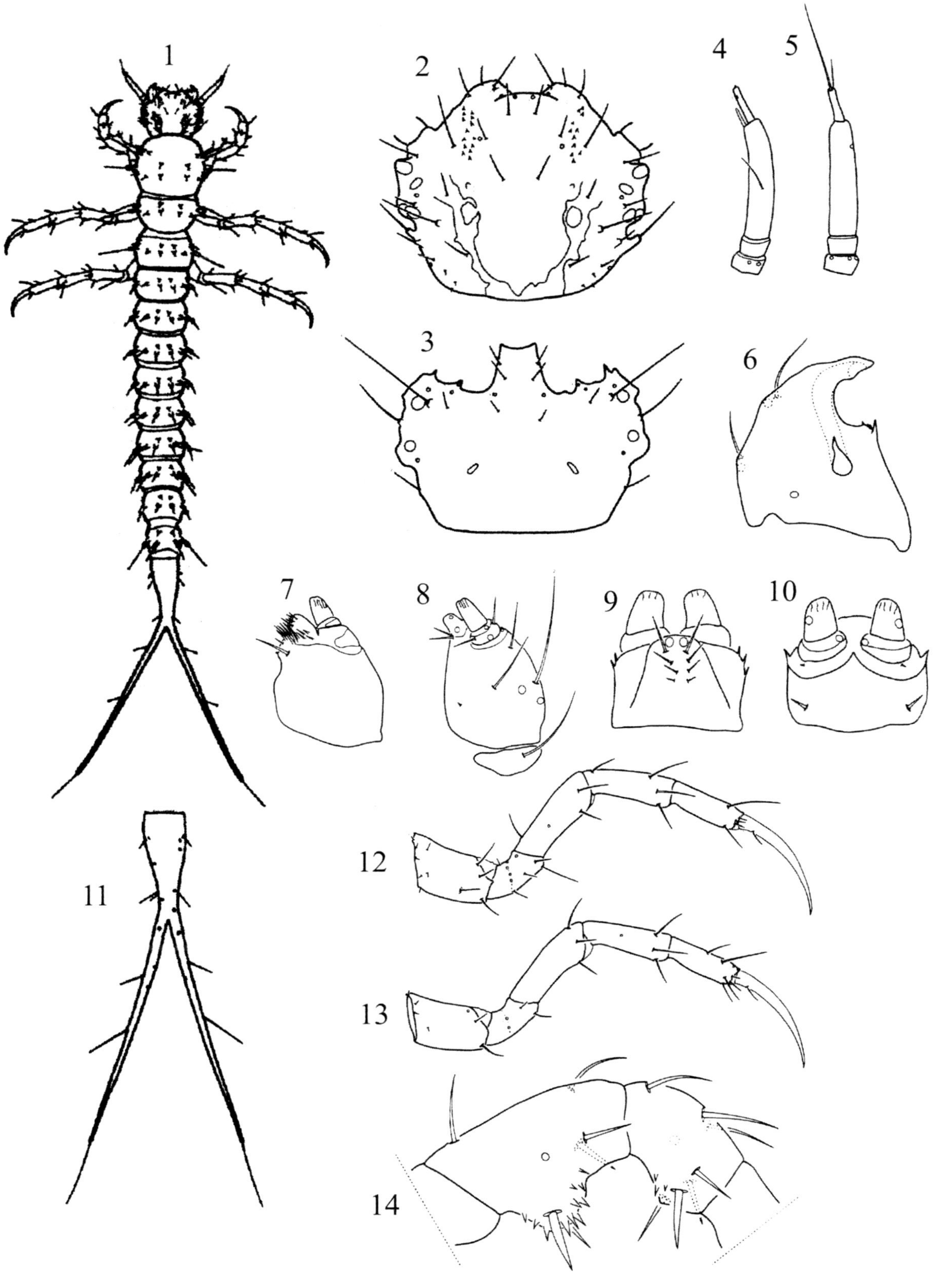

Plate 71. Figs. 1–14. *Haliplus (Liaphlus) indistinctus*. 1st instar larva. 1, larva in dorsal view; 2, head in dorsal view; 3, head in ventral view; 4, antenna in dorsal view; 5, antenna in ventral view; 6, mandible; 7, right maxilla in dorsal view; 8, right maxilla in ventral view; 9, labium in dorsal view; 10, labium in ventral view; 11, abdominal segment X; 12, left hindleg in dorsal view; 13, left hindleg in ventral view; 14, tibia and femur of left foreleg in dorsal view. All Figs. after Michat et al. (2020).

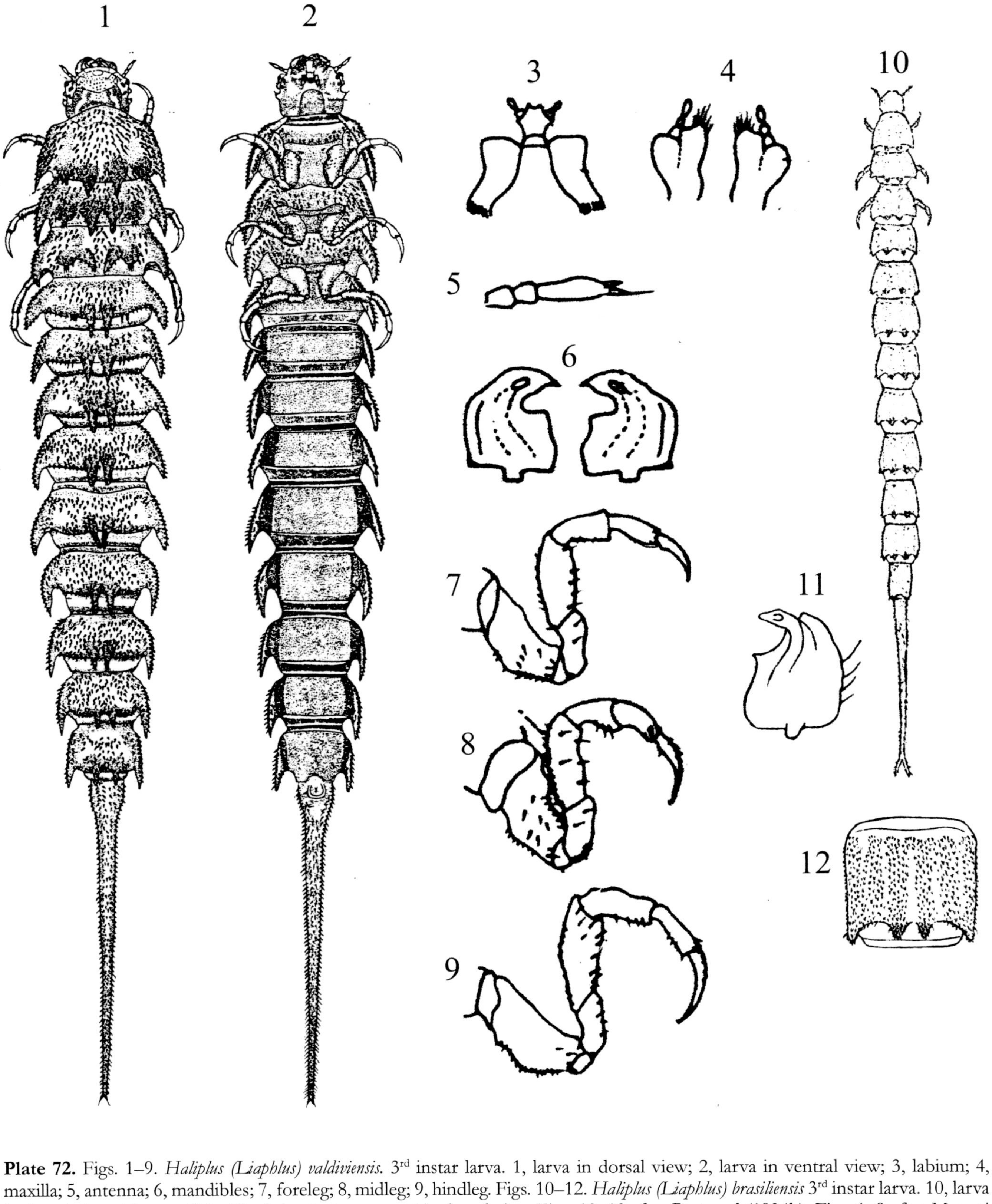

Plate 72. Figs. 1–9. *Haliplus (Liaphlus) valdiviensis.* 3rd instar larva. 1, larva in dorsal view; 2, larva in ventral view; 3, labium; 4, maxilla; 5, antenna; 6, mandibles; 7, foreleg; 8, midleg; 9, hindleg. Figs. 10–12. *Haliplus (Liaphlus) brasiliensis* 3rd instar larva. 10, larva in dorsal view; 11, mandible; 12, abdominal segment I in dorsal view. Figs. 10–12 after Bertrand (1934b). Figs. 1–9 after Moroni (1989).

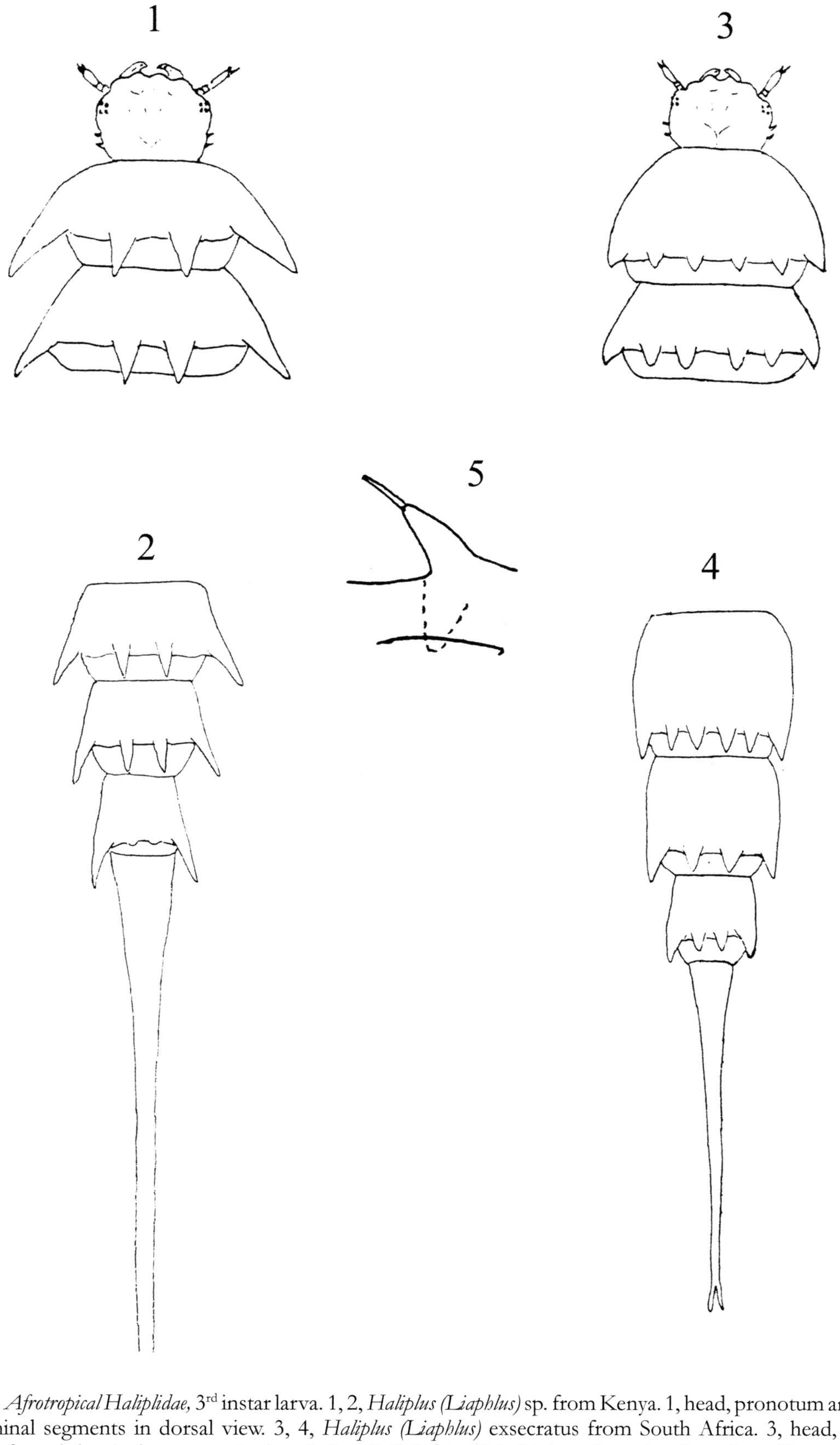

Plate 73. Figs. 1–5. *Afrotropical Haliplidae,* 3rd instar larva. 1, 2, *Haliplus (Liaphlus)* sp. from Kenya. 1, head, pronotum and mesonotum; 2, last four abdominal segments in dorsal view. 3, 4, *Haliplus (Liaphlus)* exsecratus from South Africa. 3, head, pronotum and mesonotum; 4, last four abdominal segments in dorsal view. 5, *Peltodytes (Peltodytes) quadratus*, basal tooth on filamentous dorsal and lateral process. All Figs. after Bertrand (1963).

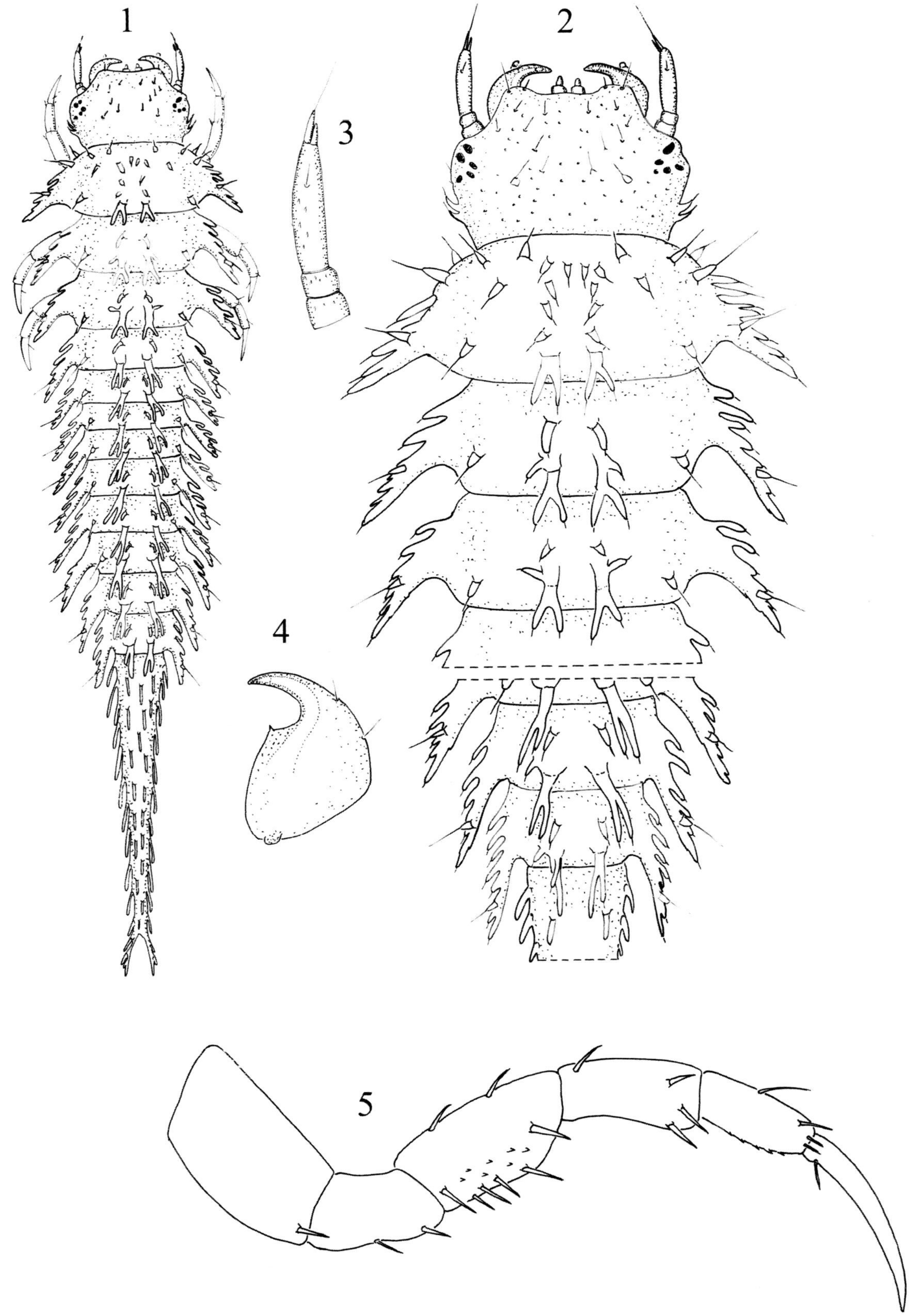

Plate 74. Figs. 1–5. *Haliplus (Liaphlus) halsei.* 1st instar larva. 1, larva in dorsal view; 2, head and thorax in dorsal view; 3, antenna; 4, mandible; 5, foreleg. All Figs. after Vondel (2012).

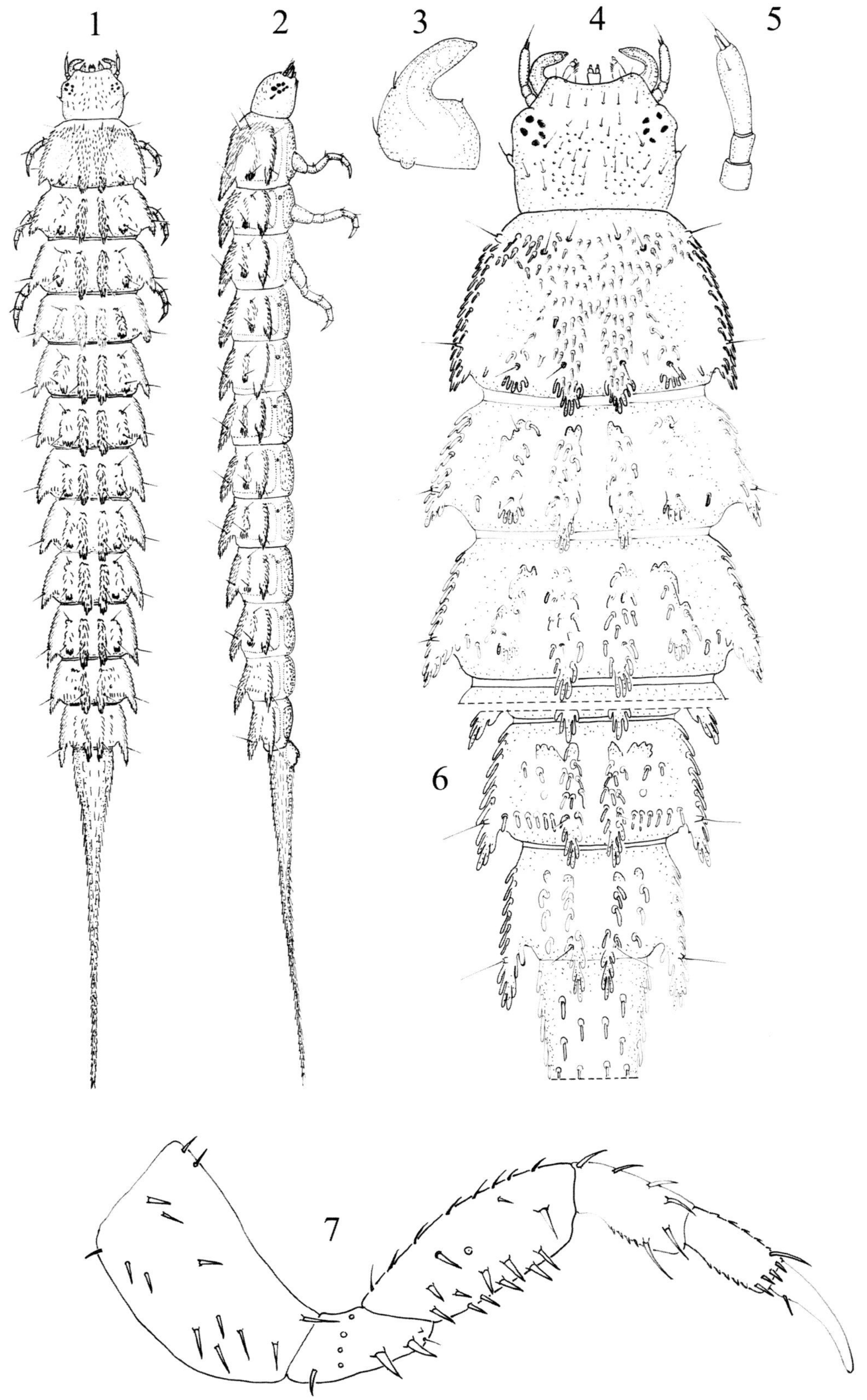

Plate 75. Figs. 1–7. *Haliplus (Liaphlus) halsei*. 3rd instar larva. 1, larva in dorsal view; 2, larva in lateral view; 3, mandible; 4, head and thorax in dorsal view; 5, antenna; 6, abdominal segment VIII, IX and base of X in dorsal view; 7, foreleg. All Figs. after Vondel (2012).

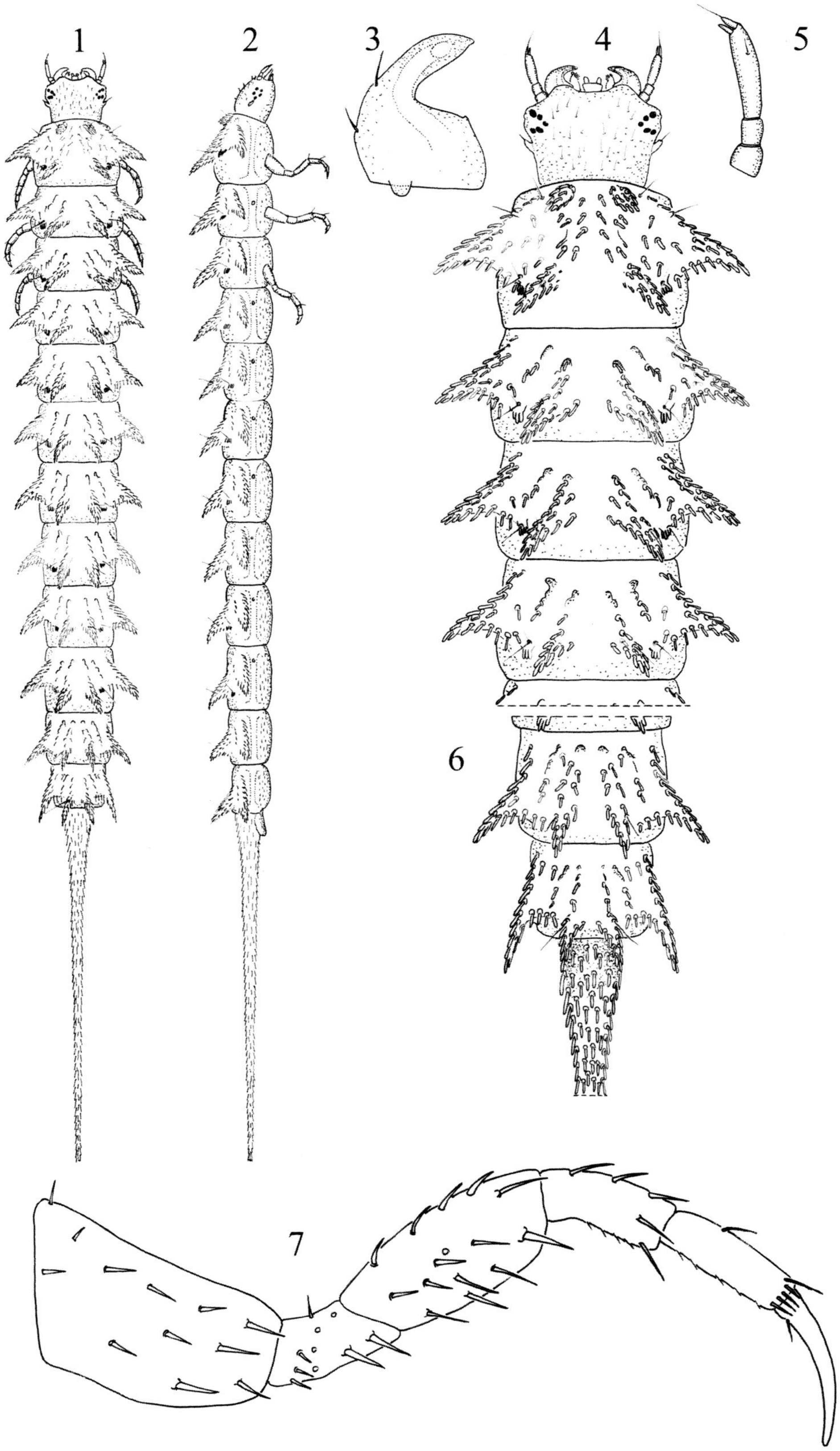

Plate 76. Figs. 1–7. *Haliplus (Liaphlus) pilbaraensis.* 3rd instar larva. 1, larva in dorsal view; 2, larva in lateral view; 3, mandible; 4, head, thorax and abdominal segment I in dorsal view; 5, antenna; 6, abdominal segment VIII, IX and base of X in dorsal view; 7, foreleg. All Figs. after Vondel (2012).

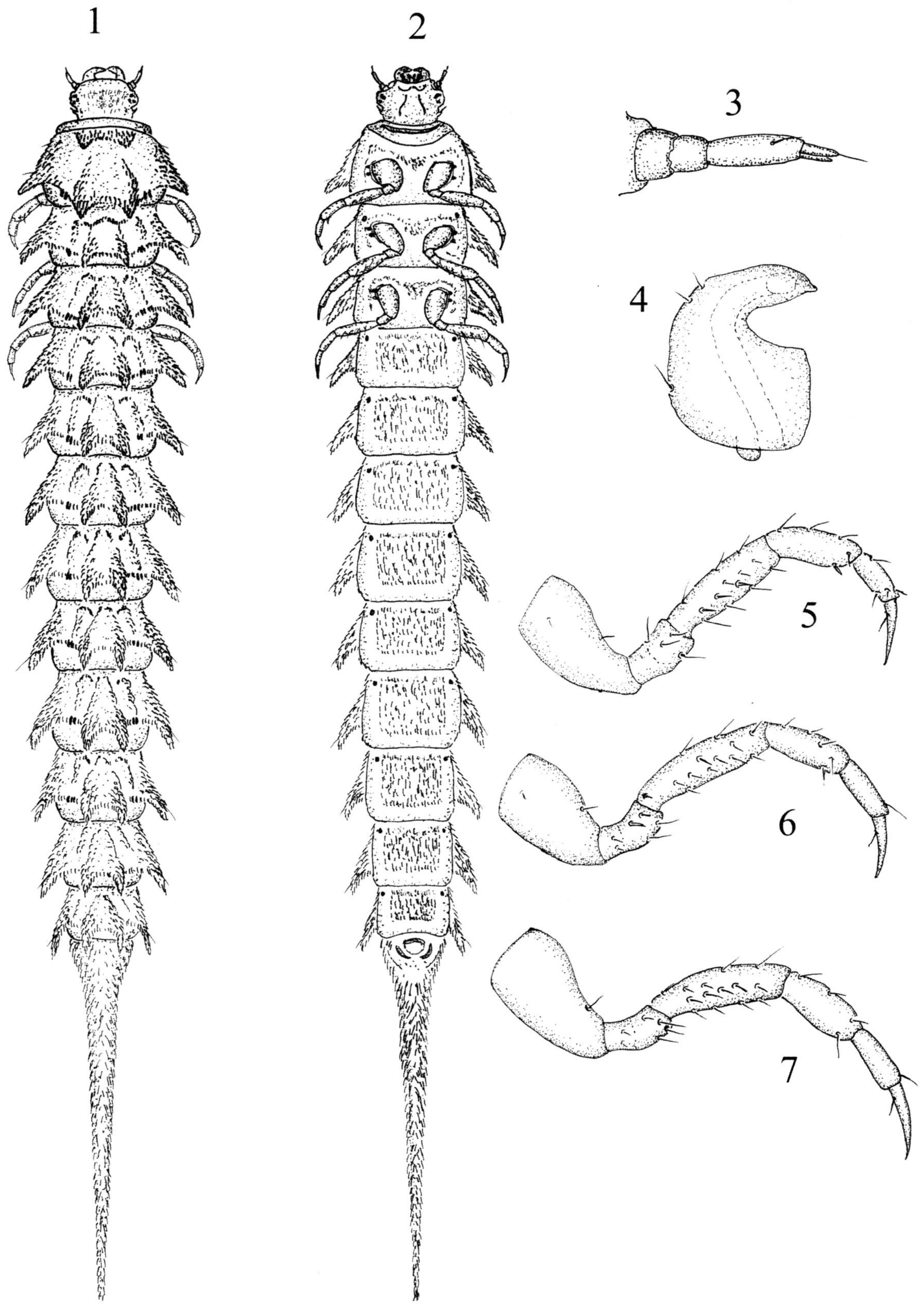

Plate 77. Figs. 1–7. *Haliplus (Liaphlus) testudo*. 3rd instar larva. 1, larva in dorsal view; 2, larva in ventral view; 3, antenna; 4, mandible; 5, foreleg; 6, midleg; 7, hindleg. All Figs. after Vondel (2004).

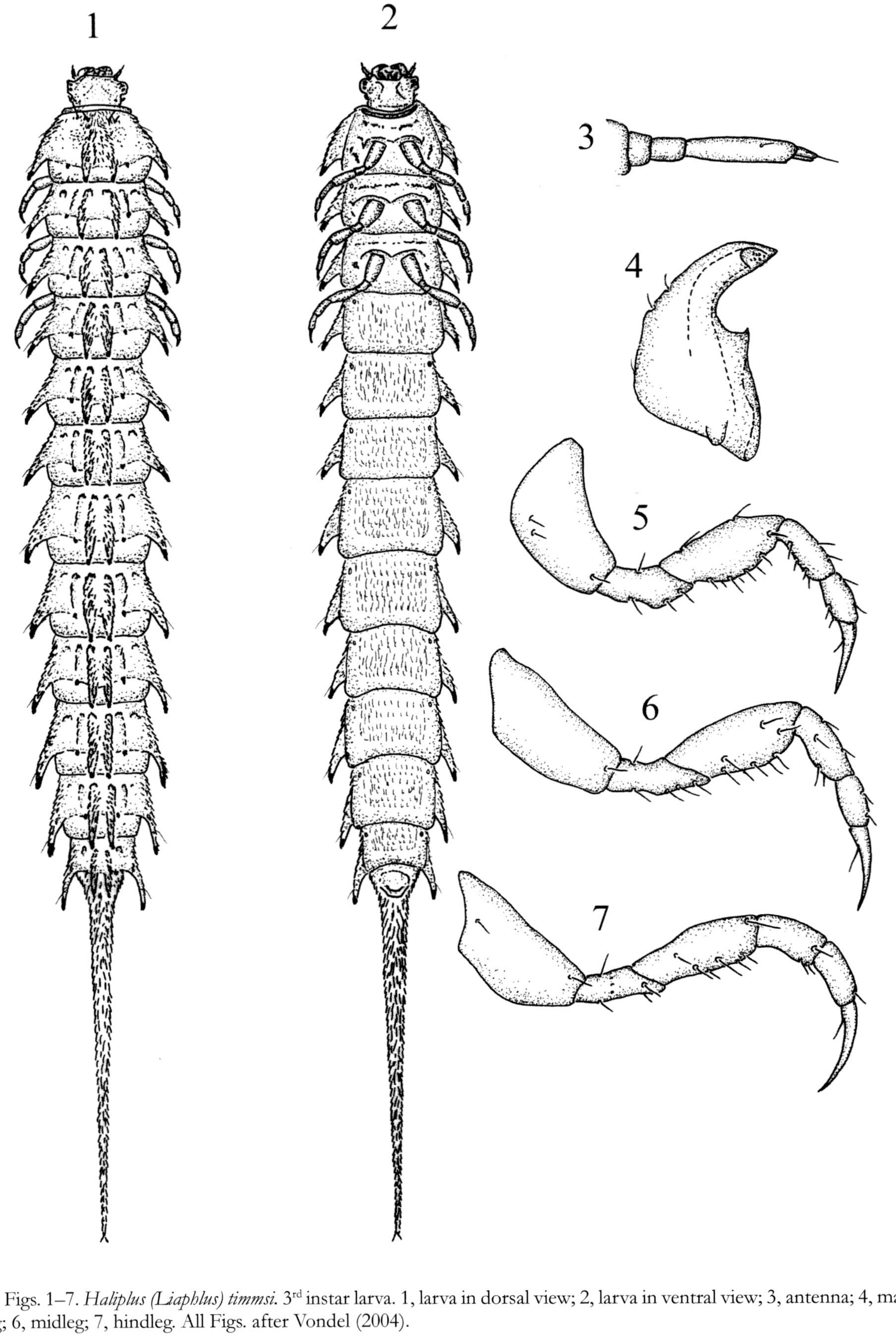

Plate 78. Figs. 1–7. *Haliplus (Liaphlus) timmsi*. 3rd instar larva. 1, larva in dorsal view; 2, larva in ventral view; 3, antenna; 4, mandible; 5, foreleg; 6, midleg; 7, hindleg. All Figs. after Vondel (2004).

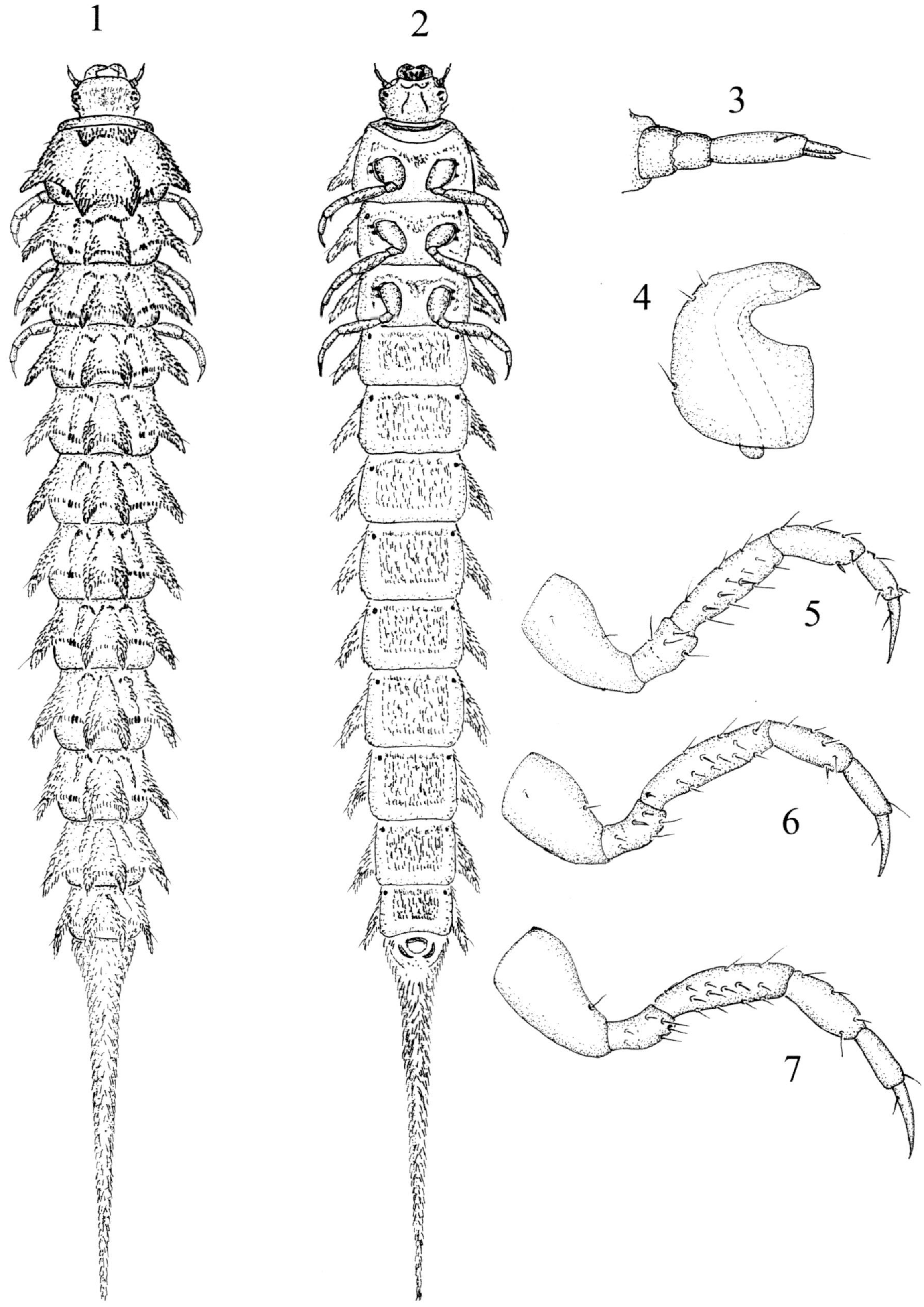

Plate 77. Figs. 1–7. *Haliplus (Liaphlus) testudo.* 3rd instar larva. 1, larva in dorsal view; 2, larva in ventral view; 3, antenna; 4, mandible; 5, foreleg; 6, midleg; 7, hindleg. All Figs. after Vondel (2004).

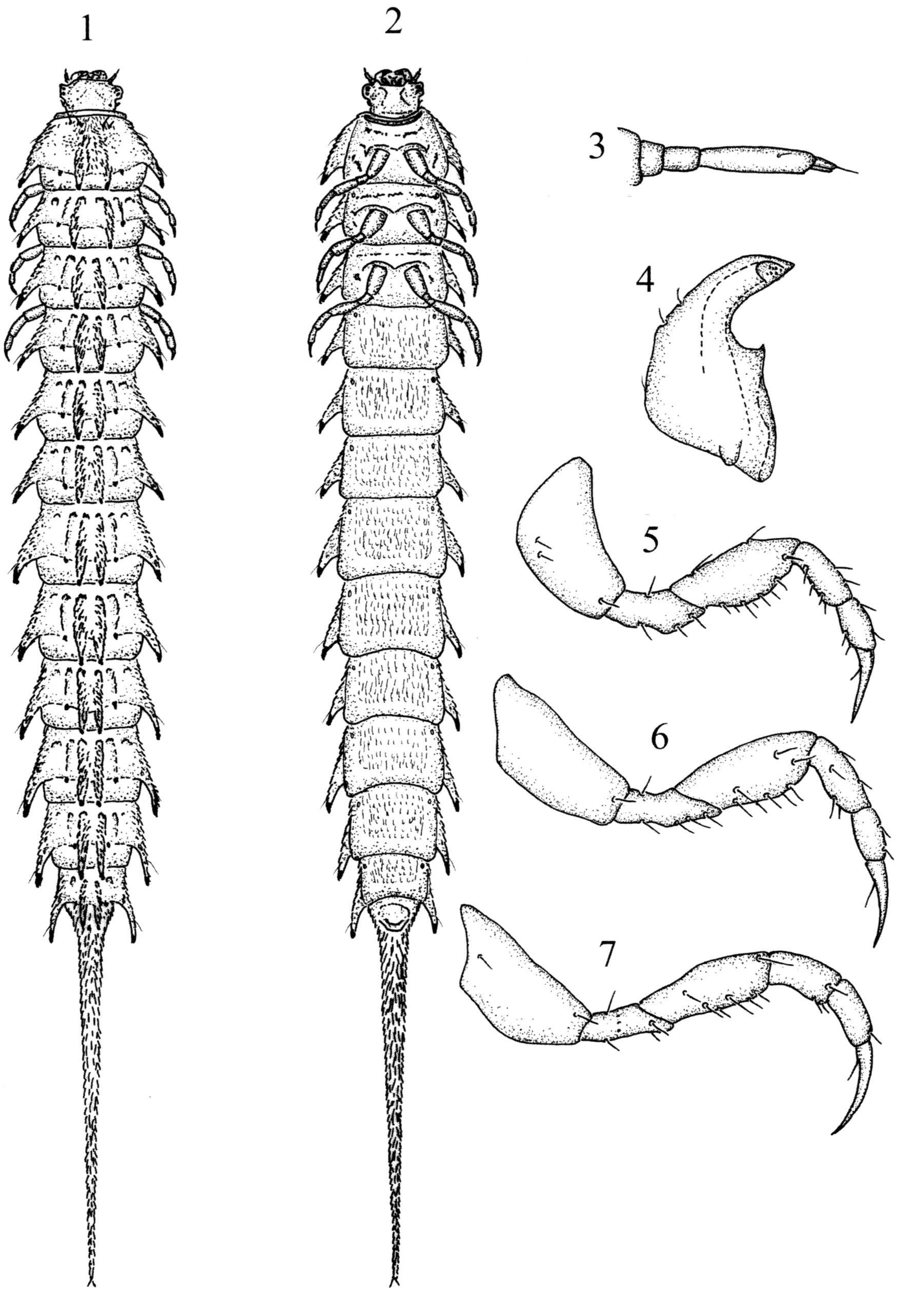

Plate 78. Figs. 1–7. *Haliplus (Liaphlus) timmsi.* 3rd instar larva. 1, larva in dorsal view; 2, larva in ventral view; 3, antenna; 4, mandible; 5, foreleg; 6, midleg; 7, hindleg. All Figs. after Vondel (2004).

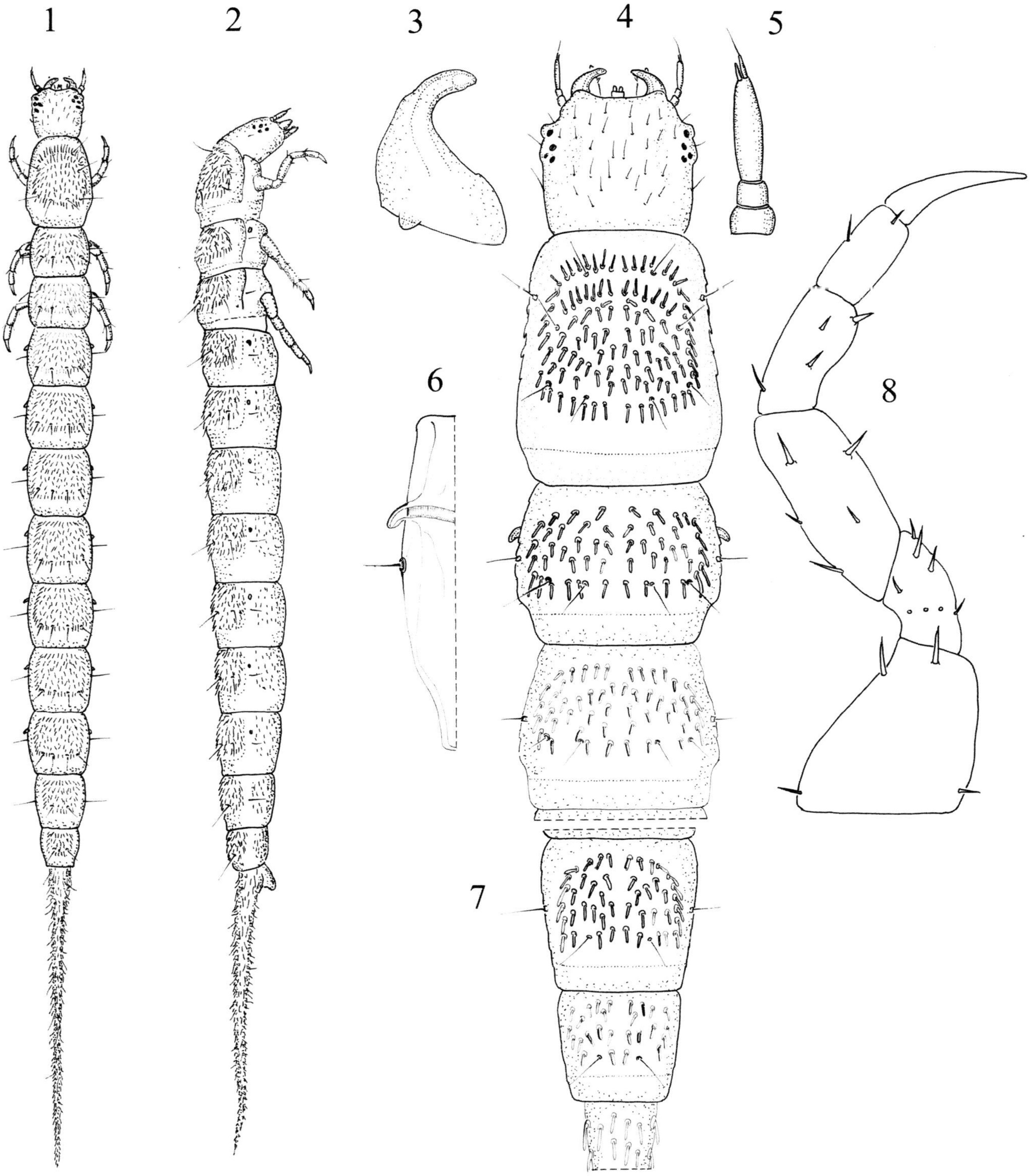

Plate 79. Figs. 1–8. *Phalilus fortescueensis.* 3rd instar larva. 1, larva in dorsal view; 2, larva in lateral view; 3, mandible; 4, head and thorax in dorsal view; 5, antenna; 6, lateral cross–section of mesonotum; 7, abdominal segment VIII, IX and base of X in dorsal view; 8, foreleg. All Figs. after Vondel (2012).

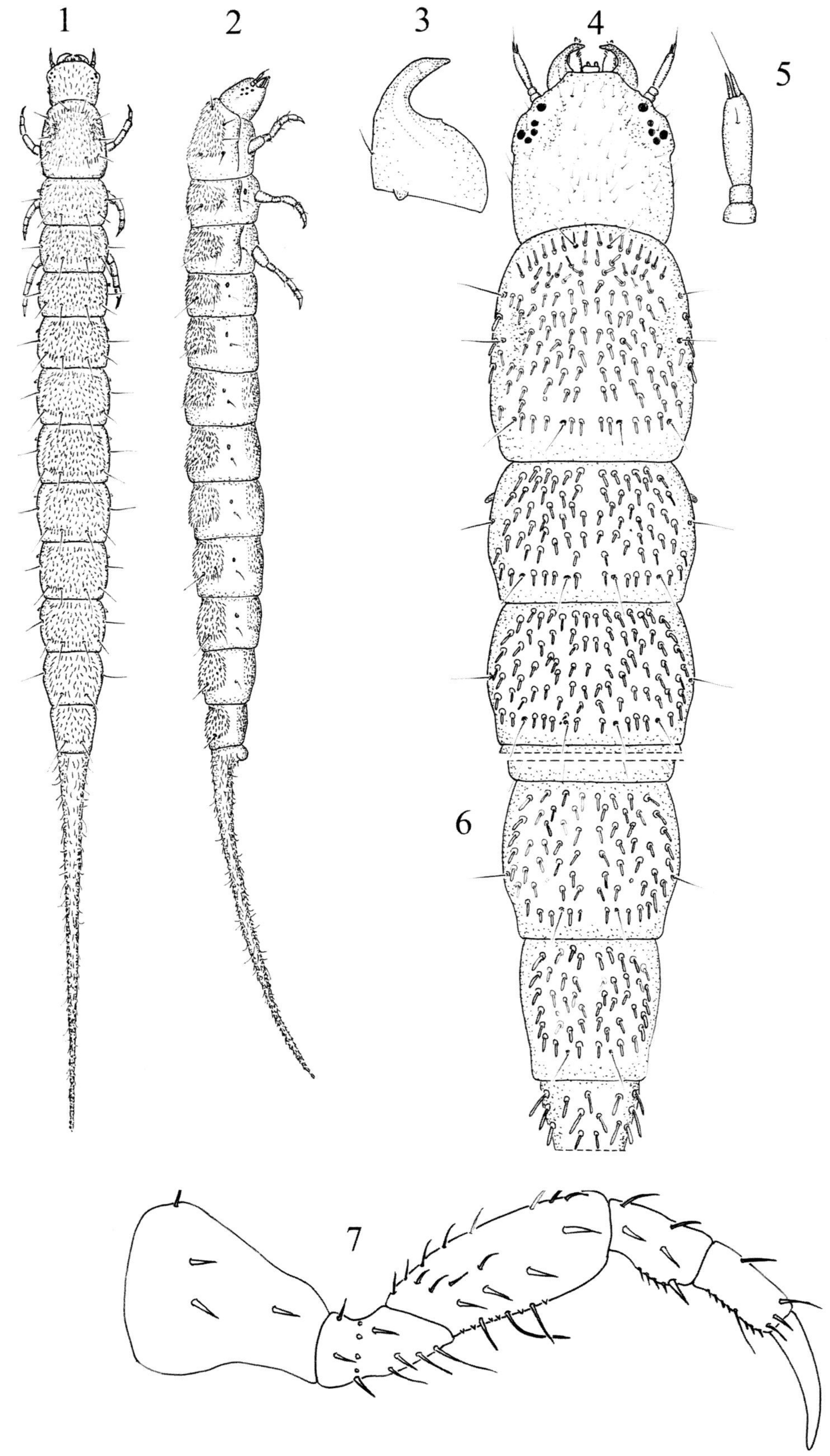

Plate 80. Figs. 1–7. *Phalilus pinderi*. 3rd instar larva. 1, larva in dorsal view; 2, larva in lateral view; 3, mandible; 4, head and thorax in dorsal view; 5, antenna; 6, abdominal segment VIII, IX and base of X in dorsal view; 7, foreleg. All Figs. after Vondel (2012).